# Development of an Infectivity Assay for Noroviruses in Cells

Awwa
Research
Foundation

## About the Awwa Research Foundation

The Awwa Research Foundation (AwwaRF) is a member-supported, international, nonprofit organization that sponsors research to enable water utilities, public health agencies, and other professionals to provide safe and affordable drinking water to consumers.

The Foundation's mission is to advance the science of water to improve the quality of life. To achieve this mission, the Foundation sponsors studies on all aspects of drinking water, including supply and resources, treatment, monitoring and analysis, distribution, management, and health effects. Funding for research is provided primarily by subscription payments from approximately 1,000 utilities, consulting firms, and manufacturers in North America and abroad. Additional funding comes from collaborative partnerships with other national and international organizations, allowing for resources to be leveraged, expertise to be shared, and broad-based knowledge to be developed and disseminated. Government funding serves as a third source of research dollars.

From its headquarters in Denver, Colorado, the Foundation's staff directs and supports the efforts of more than 800 volunteers who serve on the board of trustees and various committees. These volunteers represent many facets of the water industry, and contribute their expertise to select and monitor research studies that benefit the entire drinking water community.

The results of research are disseminated through a number of channels, including reports, the Web site, conferences, and periodicals.

For subscribers, the Foundation serves as a cooperative program in which water suppliers unite to pool their resources. By applying Foundation research findings, these water suppliers can save substantial costs and stay on the leading edge of drinking water science and technology. Since its inception, AwwaRF has supplied the water community with more than $300 million in applied research.

More information about the Foundation and how to become a subscriber is available on the Web at **www.awwarf.org.**

# Development of an Infectivity Assay for Noroviruses in Cells

Prepared by:
**Timothy M. Straub**
Battelle Memorial Institute, Pacific Northwest Division, Richland, WA 99354

**Cheryl A. Nickerson**
The Biodesign Institute, Arizona State University, Tempe, AZ 85287-5401

and

**Charles P. Gerba**
Department of Soil, Water, and Environmental Science, University of Arizona, Tucson, AZ 85721

Jointly sponsored by:
**Awwa Research Foundation**
6666 West Quincy Avenue, Denver, CO 80235-3098

and

**U.S. Environmental Protection Agency**
Washington, DC

Published by:

Awwa Research Foundation

American Water Works Association

IWA Publishing

## DISCLAIMER

This study was jointly funded by the Awwa Research Foundation (AwwaRF) and the U.S. Environmental Protection Agency (USEPA) under Cooperative Agreement No. CR-83110401. AwwaRF and USEPA assume no responsibility for the content of the research study reported in this publication or for the opinions or statements of fact expressed in the report. The mention of trade names for commercial products does not represent or imply the approval or endorsement of either AwwaRF or USEPA. This report is presented solely for informational purposes.

# CONTENTS

# TABLES

# FIGURES

# FOREWORD

The Awwa Research Foundation is a nonprofit corporation that is dedicated to the implementation of a research effort to help utilities respond to regulatory requirements and traditional high-priority concerns of the industry. The research agenda is developed through a process of consultation with subscribers and drinking water professionals. Under the umbrella of a Strategic Rcsearch Plan, the Research Advisory Council prioritizes the suggested projects based upon current and future needs, applicability, and past work; the recommendations are forwarded to the Board of Trustees for final selection. The foundation also sponsors research projects through the unsolicited proposal process; the Collaborative Research, Research Applications, and Tailored Collaboration programs; and various joint research efforts with organizations such as the U.S. Environmental Protection Agency, the U.S. Bureau of Reclamation, and the Association of California Water Agencies.

This publication is a result of one of those sponsored studies, and it is hoped that its findings will be applied in communities throughout the world. The following report serves not only as a means of communicating the results of the water industry's centralized research program but also as a tool to enlist the further support of the nonmember utilities and individuals.

Projects are managed closely from their inception to the final report by the foundation's staff and large cadre of volunteers who contribute their time and expertise. The foundation serves a planning and management function, and awards contracts to other institutions such as water utilities, universities, and engineering firms. The funding for this research effort comes primarily from the Subscription Program, through which water utilities subscribe to the research program and make an annual payment proportionate to the volume of water they deliver and consultants and manufacturers subscribe based on their annual billings. The program offers a cost-effective and fair method for funding research in the public interest.

A broad spectrum of water supply issues is addressed by the foundation's research agenda: resources, treatment and operations, distribution and storage, water quality and analysis, toxicology, economics, and management. The ultimate purpose of the coordinated effort is to assist water suppliers to provide the highest possible quality of water economically and reliably. The true benefits are realized when the results are implemented at the utility level. The foundation's trustees are pleased to offer this publication as a contribution toward that end.

David E. Rager
Chair, Board of Trustees
Awwa Research Foundation

Robert C. Renner, P.E.
Executive Director
Awwa Research Foundation

# ACKNOWLEDGMENTS

Development of this infectivity assay for human noroviruses took a tremendous effort by a number of individuals. As principal investigator, I am honored to have the following people on my team:

- Dr. Cheryl Nickerson at the Biodesign Institute, Arizona State University.
- Dr. Kerstin Honer zu Bentrup, Tulane University Medical Center.

Without Cheryl and Kerstin's expertise in the 3-D cell culture systems this project would not have been successful. Moreover, they leveraged their project finances to provide the team the cell cultures needed to make this project successful.

- Dr. Charles Gerba and Patricia Orosz Coghlan, University of Arizona. Their expertise in virological methods and their endless supply of documented norovirus stocks allowed us to examine genetically diverse strains of this virus to give us a better understanding of the potential for this assay.
- Brooke Mayer, Department of Homeland Security Fellow and graduate student at Arizona State University. Brooke performed most of the transmission electron micrographs you will see in this document. In addition she started to develop the fluorescent in-situ hybridization work that we later optimized. DHS provided Brooke's summer stipend at no charge to this project.

My team at Battelle Pacific Northwest Division:

- Rachel Bartholomew, who optimized the quantitative reverse transcription real time PCR assays.
- Alice Dohnalkova for her sample preparation and technical assistance to Brooke with the TEM work.
- Catherine Valdez for her day-to-day bench skills: PCR, cloning, sequencing, and the infectivity assays
- Cindy Bruckner-Lea for technical review of this work.

I also wish to recognize that this work would not have been possible without internal funding initiatives at our laboratory. The Environmental Biomarkers Initiative, Response to Biological Agents focus area is examining early human host biomarkers of infectious diseases. Through this initiative we received training from Dr. Nickerson and Dr. Honer zu Bentrup to develop our capabilities to generate these novel 3-D tissue assemblies to study dangerous infectious diseases, and to purchase the equipment and infrastructure needed to perform this work. Our idea was to have an integrated "omics" (genomic, proteomic, metabolomic, and volatile headspace analysis) approach to identify unique host signatures of infectious diseases. The 3-D cell culture work provided a platform to develop these tools. This jointly sponsored AwwaRF and USEPA project is a specific application of the products from this research initiative that meets a specific client need. Information on host biomarkers of response to norovirus infection is provided as a courtesy to Awwa Research Foundation and the USEPA. Drs Jon Wahl and David Wunschel are both recognized for their analysis of norovirus infected cells for unique VOC signatures and host secreted putative peptides, respectively.

Finally we wish to recognize the Awwa Research Project Manager, John Albert and members of the Project Advisory Committee for their helpful suggestions throughout this project. The project advisory committee is:

- Dr. Christine Moe, Emory University
- Jennifer Cashdollar, United States Environmental Protection Agency, Cincinnati
- Dr. John Neill, United States Department of Agriculture
- Dr. Mohammad Karim

Sincerely,

Timothy M. Straub, Ph.D

# EXECUTIVE SUMMARY

Lack of suitable tissue culture methods hampers the ability to determine the viability and infectivity of certain United States Environmental Protection Agency (USEPA) Contaminant Candidate List (CCL) pathogens isolated in source and drinking water supplies. In addition to gathering occurrence data in water supplies, these infectivity assays assist water utilities and USEPA in determining the risk of infection should these pathogens be detected by nucleic acid methods (e.g. polymerase chain reaction (PCR) and its variant, quantitative real-time PCR) in public water supplies.

Both *in vitro* infectivity and cultivation assays have proven problematic for the waterborne CCL pathogen, human norovirus (NoV), since its recognition over 30 years ago. Recent advances in engineering *in vitro* 3-dimensional tissue assemblies that recapitulate many of the physiological features of their *in vivo* counterparts is the central enabling technological platform that allowed us to overcome the NoV cell culture infectivity problem. Detailed methods for propagating these novel tissue assemblies and results from infectivity trials are provided in this report. With training and technology transfer, large water utility, and state health laboratories with functioning waterborne virology capabilities will likely be able to replicate these methods. While the methods for generating these novel assemblies are complex, the infectivity assays are fairly straightforward (e.g. standard $TCID_{50}$ assays and reverse-transcription real time PCR). This research advance will now allow for testing disinfectants and UV for their efficacy to control NoV in municipal drinking water supplies.

## RESEARCH OBJECTIVE

The primary objective for this work was to develop an infectivity assay for NoV using *in vitro* cell culture techniques. To date, many different cell lines, co-cultures of cell lines, and media supplements have been used to initiate infection, but no success with human NoV has been achieved. Our central hypothesis was that *in vitro* cell culture models used to study host-pathogen interactions have benefited from the recognition that organs and tissues function in a 3-dimensional environment and that this spatial context is necessary for development of cultures that more realistically resemble *in vivo* tissues and organs from which they were derived. This may be particularly the case for NoV, and thus we pursued novel 3-D cell culture technology to achieve our objective. A number of techniques were used to demonstrate NoV infectivity, and these are described in the research approach below.

## REASEARCH APPROACH

### Cell Culture Methods

We chose the small intestinal cell line INT-407 (ATCC CCL-6) to develop the infectivity assay for NoV. These cells were grown as monolayers in GTSF-2 tissue culture media in standard T75 tissue culture flasks, and incubated at 37°C and 5% $CO_2$ until they reached 70% confluency. Cells were then harvested (5 x $10^6$ total cells) from these flasks and added to porous collagen-I coated microspheres. Upon colonization, the bead-cell complexes were transferred to a rotating wall vessel (RWV) bioreactor. The RWV bioreactor was filled with GTSF-2 media and the rotation speed was initially set at 18 RPM, based on previous optimization with this cell line model in this system. These conditions mimic the physiological fluid shear these cells experience *in vivo*. Rotation speed was adjusted throughout the course of the incubation period to maintain these aggregates in a constant state of free-fall. Media was exchanged every 24-72 hrs, depending on the age of the cells in the bioreactor, with the more frequent exchanges occurring at the latter stages of incubation. After 35 days of growth and differentiation, the tissue aggregates were transferred to 24-well tissue culture dishes to perform infectivity assays. Cell batches were tested for markers of cellular differentiation to ensure quality control (QC). Correct localization of these markers indicated successful differentiation.

Dilutions of norovirus stocks from stool samples from documented outbreaks were used to perform initial infections. Viruses harvested from these initial infections were subsequently passed in cell culture to demonstrate continued infectivity and propagation of the viruses. Negative controls consisted of NoV negative stool samples and ultra-filtered (10,000 dalton) positive stool and tissue culture lysates. The virus-free samples were then applied to cells and assayed as if they were NoV positive samples.

**Assays Performed to Demonstrate Infectivity**

Limiting dilution reverse transcription PCR (RT-PCR) was performed to qualitatively determine if viral replication was occurring. Primer sequences amplifying the RNA-dependent RNA polymerase gene were used for most PCR assays, and these products were cloned and sequenced to determine if these viruses retained their same sequence as the original outbreak sample, or if potential sequence changes were occurring, potentially indicating adaptation of the viruses to these cell cultures. Towards the conclusion of this project, reverse transcription real-time PCR was optimized to characterize productive viral replication in cell culture.

Fluorescent *in situ* hybridization (FISH) assays targeting internalized NoV RNA were performed in two of the 4 infection trials described in this report. These probes targeted the highly conserved RNA sequences of the ORF1-ORF2 junction of NoV. In one infection trial, biotin-labeled probes were used, but background problems made data interpretation difficult. In the subsequent infection trial, the chemistry of these probes was switched to molecular beacon (MB FISH) probes. For this latter set of probes, the MB FISH probes only generate signal when the correct RNA sequence is present, thus mitigating background fluorescence problems.

Microscopic techniques included examining the gross pathology of the infected tissue aggregates using an inverted microscope and sub-cellular pathology determined by thin- and ultra-thin section light microscopy (LM) and transmission electron microscopy (TEM), respectively. In the case of LM, a QC check determined if numerous, well-formed microvilli were present on the apical surfaces of the 3-D small intestinal cells.

**RESULTS AND DISCUSSION**

Of the infection trials described in our published study in *Emerging Infectious Diseases* (Straub et al. 2007), we report that we have successfully developed the first *in vitro* cell culture infectivity assay for noroviruses. Uninfected cells appeared as uniform monolayers attached to their microcarrier beads, but for infected tissue aggregates, cells lost contact with their microcarrier beads, and sloughed as large mucoid mats from the beads (Figure E1, Panel A). Sub-cellular pathology revealed significant vacuolization, shortening of the microvilli, internal rearrangement of organelles, and accumulation of suspect particles that are the correct size and shape for NoV (Figure E1, Panel B). Limiting dilution PCR provided evidence that NoV replication was occurring (Table E1). We believe that this was a true increase in viral RNA, and not an evaporation artifact, as media was changed in these aggregates every 24 hrs. Because the aggregates are not attached to the plate, not all of the media can be removed from these aggregates, and thus a dilution, not a concentration effect should have been observed, if viral RNA replication were not occurring. In addition, MB-FISH revealed viral RNA inside the cells (Figure E1, Panel C).

Additional work not published in the journal article provides additional lines of evidence that human noroviruses can infect these 3-D cell culture models. Reverse transcription real-time PCR was optimized to characterize the level of replication in cell culture. Norovirus GI.1 provided from a human challenge study, and additional GII isolates from CDC investigated outbreaks were compared, and these results are summarized in Table E2 below. In general, the degree of viral replication was variable (approximately >10 fold to over 1,000 fold), but statistically significant. For higher dose infection, maximum viral replication was observed at 24 hrs post infection (pi), whereas for low dose infections, significant viral replication was observed at 72 – 96 hrs pi. Additional work should be conducted to determine address whether this variability in viral replication is due to the assay method or strain.

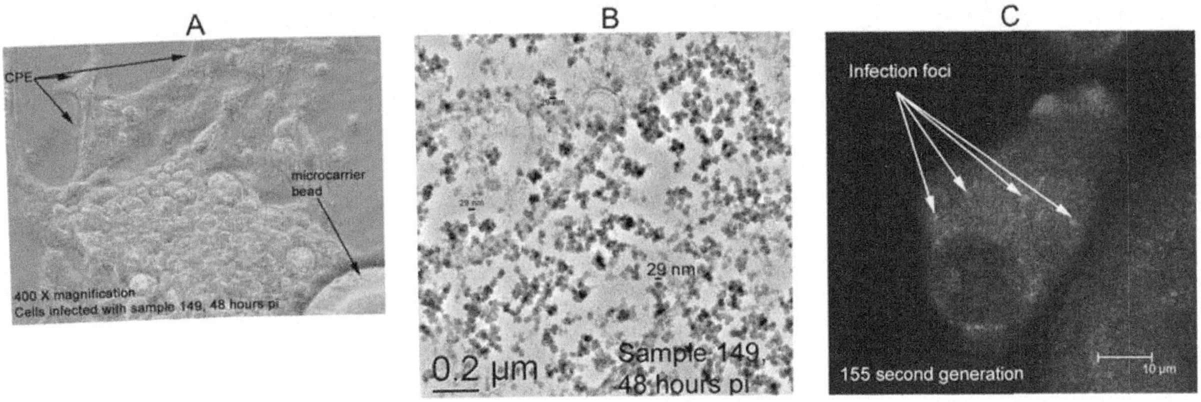


**Figure E1 Evidence of NoV infection. A) Gross cytopathic effect (CPE), uninfected cells remain attached as monolayers to the microcarrier beads, B) TEM showing numerous suspect particles and internal membrane rearrangement, C) MB-FISH showing viral RNA foci inside the cells.**

**Table E1**

**Relative increases in viral RNA in infected tissue aggregates as measured by end-point dilution PCR**

| | Hours post infection | | | |
|---|---|---|---|---|
| Effective dilution of working viral stock applied to cells | 1 hr | 24 hr | 66 hr | 72 hr |
| 1:10 | + | + | + | Not done |
| 1:100 | + | + | + | Not done |
| 1:1000 | - | - | + | + |
| Negative control | - | - | - | - |

**Table E2**

**Reverse transcription, real-time PCR data for NoV infections of cell cultures: supernatant fraction**

| Sample ID | Genogroup | Mean copies applied to cell culture* | Mean copies per mL recovered 24 hrs pi | 48 hrs pi | 72 hrs pi | 96 hrs pi | Setting |
|---|---|---|---|---|---|---|---|
| VS 122 | II.4 | < Limit of detection (LOD)† | < LOD | 1,922 | 7,075 | ND‡ | Cruise ship |
| flag1 | II | < LOD | < LOD | < LOD | < LOD | ND | Nursing home |
| Patient 3-1 | I.1 | < LOD | < LOD | < LOD | < LOD | ND | Clinical pre-challenge |
| Patient 3-9 | I.1 | 80,200 | 989,703 | 36,143 | < LOD | ND | Clinical post challenge |
| Patient 4-1 | I.1 | < LOD | < LOD | < LOD | < LOD | ND | Clinical pre-challenge |
| Patient 4-3 | I.1 | 6,960 | 6,840,255 | 165,335 | 396,062 | ND | Clinical post challenge |
| Patient 4-1§ | I.1 | < LOD | < LOD | < LOD | < LOD | ND | Clinical pre-challenge |
| Patient 4-3§ | I.1 | 96,500 | 21,685,800 | 5,536,800 | 1,007,390 | 422,950 | Clinical post challenge |
| 127** | II.4 | <LOD | <LOD | <LOD | <LOD | <LOD | Cruise ship |
| 8G | II | 457,800 | 26,684,300 | 5,767,500 | 1,540,307 | 464,476 | Cruise ship |

* Viruses in stools were diluted in tissue culture media and then applied to cells. This represents the intrinsic copies applied to the cells, prior to being overlain with 1 mL of tissue culture media.

† The LOD for our assay was consistently between 100 and 1,000 reverse transcription real-time PCR copies per mL.

‡ ND = Not performed for this timepoint.

§ This is a repeated experiment with a different batch of cells and on a different date. Note that 3-1 and 4-1 are stool samples from two different patients that are not infected with norovirus (e.g. stool sample negative control).

** Viral replication was observed in the cell fraction at 96 hrs post infection (26,915 copies/mL).

Finally, collaborative efforts by other investigators at Pacific Northwest National Laboratory revealed that the host cells are responding to viral challenge. "Shotgun proteomics" revealed putative up-regulated peptides in response to viral challenge as compared to controls (host secreted peptides observed in infected cells vs. uninfected cells) and putative down regulated peptides (peptides observed in the controls, but absent in infected cells) (Figure E2). Headspace analysis of metabolic volatile organic compounds from infected cells also revealed the presence of unique chemicals emitted from infected cells that were absent in controls.

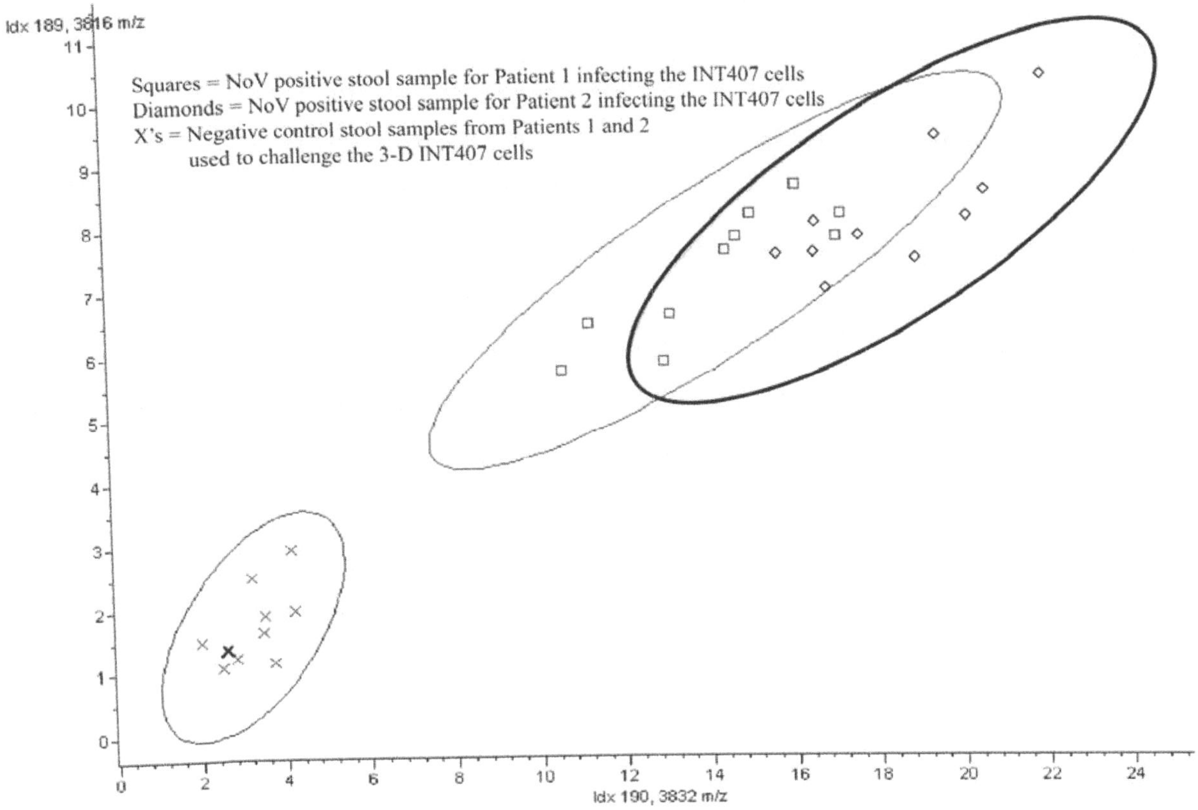


**Figure E2 1-D MALDI-MS analysis showing significant differences in host-secreted peptides between uninfected (indicated by the x's) and GI.1 NoV infected cell cultures (indicated by the squares and diamonds (patient samples 3-9, and 4-3)). In this graph, the software picked two mass/charge ratio (m/z) peaks that appeared to be significantly different between uninfected and infected cells, respectively. The axes represent the relative peak areas for a given sample and m/z ratio. Ovals around the indicated peak areas represent the confidence interval for the respective samples. Non-overlapping areas are statistically significant. This data was provided as a courtesy from Dr. David Wunschel Pacific PNNL.**

## CONCLUSIONS

The work presented in this report lays the foundation for other studies that should be conducted. For all future work, it is imperative that if viral challenge studies are going to be conducted with these 3-D aggregates, the cells must exhibit markers of cellular differentiation and formation of numerous well-formed microvilli. Experiments relevant to the water industry can now be performed. These would include assays to test efficacy of chemical disinfectants (e.g. chlorine, chloramines, and ozone), and UV irradiation. Other research may include gathering true occurrence data in groundwater, surface source water, and finished drinking water to better assess risks from consumption of potentially contaminated drinking water.

## FOOTNOTE

A major portion of the work discussed here has been published in the United States Centers for Disease Control Journal *Emerging Infectious Diseases* 13(3) 396-403. The authors retain the copyright, and we are free to quote the work, provided that appropriate citations are used. The complete reference is provided below:

Straub, T. M., K. Honer zu Bentrup, P. Orosz-Coghlan, A. Dohnalkova, B. K. Mayer, R. Bartholomew, C. O. Valdez, C. J. Bruckner-Lea, C. P. Gerba, M. Abbaszadegan and C. A. Nickerson 2007. I*n vitro* Cell Culture Assay for Human Noroviruses. *Emerg. Infect. Dis*. 13(3): 396-403.

# CHAPTER 1 INTRODUCTION

## STATEMENT OF THE PROBLEM

Noroviruses (NoV) are non-enveloped positive sense, single stranded RNA viruses that are approximately 27 to 35 nm in diameter (Caul et al. 1982; Kapikian et al. 1972). They belong to the genus *Norovirus* in the *Caliciviridae* family and consist of three genogroups (I, II, and IV) that infect humans (Green et al. 2000; Green et al. 2002; Vinje et al. 2004; Zheng et al. 2006). Based on sequence diversity of the capsid gene, NoV can be classified in 8 genetic clusters within GI, 17 within GII, and 1 in GIV (Zheng et al. 2006).

These viruses are the leading cause of non-bacterial, self-limiting, gastrointestinal illness worldwide (Atmar et al. 2001; de Wit et al. 2001; Hale et al. 2000; Lindesmith et al. 2003). Infected persons may develop symptoms of severe nausea, vomiting, and watery diarrhea within 12 - 36 hrs of exposure, and typically remain symptomatic for 1 – 2 days (CDC MMWR 2001; 50 (No. RR-9)). Infections may lead to fatalities in immunocompromised individuals. The most important routes of norovirus transmission are via contaminated water and food, and secondary person-to-person spread is significant (CDC MMWR 2001; 50 (No. RR-9)).

Due to the importance of this family of viruses, *Caliciviridae* are listed on USEPA's Contaminant Candidate List (CCL 2). The United States EPA uses this list of unregulated contaminants to prioritize research and data collection efforts to help them determine whether they should regulate a specific contaminant. Research on NoV occurrence and risks in municipal drinking water supplies has relied on reverse transcription real-time PCR (Jothikumar et al. 2005; Kageyama et al. 2003). However, real-time RT-PCR methods may not provide an accurate assessment of the infectivity potential for these viruses. This is because PCR only requires that the target nucleic acid sequences remain intact, regardless of whether the virions are damaged to the extent that they can no longer attach to, enter in, and replicate in suitable host cells or people. In the case of real-time PCR, only 70 – 100 bp of an intact stretch of a 7,000 nt length virus are needed for successful amplification (represents < 2% of the genome). While this method can provide pathogen occurrence data to USEPA and the water industry, viral infectivity assays provide "gold standard" evidence regarding the public health risks associated with pathogen detection in water supplies. For NoV research, the problem has always been the lack of a suitable viral infectivity assay either in cell culture or animal models.

## DEVELOPMENT OF NOV CELL CULTURE METHODS – PAST HISTORY

Since its recognition over 30 years ago, the human NoV cell culture problem has been examined in a number of seemingly unrelated research reports.

- "Shotgun" approaches to try multiple cell lines, and multiple cell treatments to assist cellular differentiation (Duizer et al. 2004; Malik et al. 2005).
- Development of plaque assays for a genetically similar NoV genus that infects mice or related *Caliciviridae* members (Green et al. 2002; Parwani et al. 1991; Wobus et al. 2004; Wobus et al. 2006).

- Potential creation of NoV challenge pools using "synthetic biology" techniques (Asanaka et al. 2005; Chang et al. 2006).
- Surrogate animal models (Cheetham et al. 2006; Rockx et al. 2005; Subekti et al. 2002).
- Characterization of recombinant NoV particles and their attachment to host cell surface expressed antigens and/or NoV protein interactions with cells (Fernandez-Vega et al. 2004; Hale et al. 2000; Harrington et al. 2002; Harrington et al. 2004; Hutson et al. 2003; Lindesmith et al. 2003; Marionneau et al. 2002; White et al. 1996).

Each of these research focus areas are reviewed below to provide the basis for the development of the NoV infectivity assay reported here.

**"Shotgun" Approaches**

Cultured mammalian cells, including primary cultures, organ cultures and continuous cells lines, have long been used to investigate host-pathogen interactions that lead to infection. While many of these methods have worked for numerous bacterial and viral studies, human noroviruses have not been successfully cultured *in vitro*. Duizer et al., (2004) published a comprehensive study that detailed efforts to propagate noroviruses in cell culture. Eight human gastrointestinal epithelial cell lines, eight cell lines derived from other human tissues, four cell lines derived from monkey kidney epithelia and seven cell lines derived from other animal tissues were evaluated. Treatments including co-culture with CCD-18 cells, dimethylsulfoxyde (DMSO), butyric acid, insulin and dexamethasone were all attempted, in order to achieve better cellular differentiation. Other methods, such as incubation in rotating vessels and on a rocking platform, hypotonic shock, centrifugation, passaging inoculated cells and incubation at lower temperatures, were also used to enhance virus contact with cells, virus internalization, and virus replication. Based on the hypothesis that biochemistry of the gut may trigger cell signaling pathways, thus allowing cells to become permissive to viral infection (Parwani et al. 1991), further additives, such as saliva, trypsin, pancreatin, human duodenal filtrate, human ileal filtrate and monkey kidney intestinal preparation were added in an attempt to mimic the conditions of the gastrointestinal tract. Other studies by Malik et al. (2005), evaluated 19 cell types inoculated with human noroviruses for five passages, but they were also unsuccessful since both cytopathic effect (CPE) and reverse transcription polymerase chain reaction (RT-PCR) had failed.

**Plaque Assays with Caliciviruses Infecting Other Animals**

Recently, the first plaque assay for a norovirus genera that infects mice was reported (Wobus et al. 2004; Wobus et al. 2006). The significance of the Wobus study was that these noroviruses have an affinity for macrophages and dendritic cells, and this provided a potential clue for development of the human norovirus cell culture. This group has recently reported that colonization of the gastrointestinal tract of the mice does occur, but these animals do not develop gastrointestinal symptoms of diarrhea or an emetic response to infection. Rather, this genogroup of norovirus appears to spread systemically in these animals. Nonetheless, because this virus can survive gastric pH, and is amenable to plaque assays, Cannon et al., found that these viruses may be better surrogates to study survival of the norovirus genus in environmental samples (Cannon et al. 2006).

Prior to the Wobus studies, other Caliciviruses have been proposed as surrogates for the study of human NoV. Parwani et al. (Parwani et al. 1991) studied porcine enteric calicivirus and found that ileal filtrates were necessary for their serial propagation in cell culture. This study likely was the motivation behind the Duizer study attempts to use intestinal content preparations to enhance the permissiveness of the cell lines they challenged with human noroviruses. Green et al., (2002) and others have also studied feline calicivirus as a human NoV surrogate as well. These viruses, however, infect lung tissue in cats. In addition, they do not exhibit pH resistance like the murine noroviruses (Cannon et al. 2006), and thus may not be suitable surrogates for human NoV infection.

**Synthetic Biology**

Several reports have surfaced regarding the production of functioning Norwalk-like virus particles using recombinant DNA techniques (Asanaka et al. 2005; Chang et al. 2006). Essentially, these techniques employ the generation of full-length cDNA from the virus. Cells are then transfected with the cDNA using a modified vaccinia virus or other suitable vector. The cDNA is essentially a "negative strand" of the RNA. New RNA for structural proteins and nucleic acid sequences for the new virions are produced. There are several reasons for employing the approaches reported by these groups. First, both groups are showing that once functioning RNA is inside the cells, the remainder of the replication cycle can occur. This implies that the primary block to natural human NoV infection in cells is likely due to attachment to, entry into, and/or unpackaging of the virions once inside the cells. Second, these new methods may be able to provide insights into pathogenesis of NoV. Third, the primary research driver for these groups is to create challenge pools of virus that they can use to test candidate vaccines, adjuvants, and therapeutics. Previously, the only way to create a challenge pool was to infect susceptible individuals, harvest and purify the viruses, and then conduct separate human trials (e.g. it is likely the group used to create the challenge pool of virus cannot be readily infected again (Lindesmith et al. 2003)). However, for the purposes of environmental monitoring, these methods are synthetic, and still do not reveal infectivity potential of the native virions.

**Animal Models**

Recent work by Cheetham (Cheetham et al. 2006) has bolstered hopes for a suitable animal model to study human NoV infectivity and pathogenesis. Gnotobiotic piglets can be infected with GII.4 human NoV. Intestinal lesions, diarrhea, and shedding of virus particles are observed. Their studies are continuing to examine the role of histo-blood group compatibility and its correlation to infectivity potential (Cheetham et al. 2007). Like humans, there is an apparent correlation with blood type and susceptibility to infection.

Experimental infection of primates has also met with limited success, perhaps due to limited numbers of subjects in the study. Rockx et al. (2005) found that 1 of 3 rhesus macaques tested could potentially be infected and shed virus for up to 19 days. The infected macaque could raise IgG and IgM, but not IgA antibodies. Subekti et al. (2002) was able to successfully infect infant pigtail macaques with the Toronto virus. Their study showed that 2 of 3 infant macaques tested were infected and exhibited clinical symptoms. When these viruses were harvested and used for a second challenge of naïve macaques, they caused clinical infection in 1 of the 2 infant macaques tested in the second group. Two of the infant mothers, one in each group, also became infected and exhibited clinical symptoms of NoV infection. While both primate models show promise for understanding human pathogenesis, both of the primate studies used a limited number of observations, and did not address mechanisms of NoV pathogenesis *per se*. Nor did either study address susceptibility to antigenically diverse members of the NoV genus.

### Histo-Blood Group Antigen Studies

Several groups have investigated the role of cell surface receptors for the attachment of human NoV. Infectivity is usually based on initial binding of the virus to an appropriate host cell receptor. Recent work has identified ABH histo-blood group antigens as ligands for the attachment of noroviruses to mucosal cell surfaces. Lindesmith et al. (2003) showed that approximately 29% of their study population did not express the H type 1 oligosaccharide and were recessive for the FUT2 gene. These individuals could not be infected with norovirus, regardless of dose. These investigators also showed that some volunteers that had the required ligand and gene were also resistant to infection, possibly indicating a memory immune response or another unidentified factor. Harrington et al. (2002) further confirmed that antisera from previously infected humans will block binding of H type 1, H type 3, and $Le^b$ antigens on these cells. Further evidence of the involvement of histo-blood group antigens in binding recombinant NoV (rNoV) particles was provided by Hutson et al. (2003). In addition to the H type 1 ligand, H type 2 is involved in hemagglutination for rNoVs for human type O RBCs. Using real noroviruses, Harrington et al. (2004), further confirmed these findings. In their study H type1 and H type 3 antigens were shown to bind NoV as well as involvement with $Lewis^a$ and $Lewis^b$ antigens.

### Information Synthesis Leading to Model Development

Synthesis of all of this information led to the development of the infectivity assay reported here. From the Duizer group attempts, it was evident that "shotgun" approaches to attempt other cell lines, chemical, and physical treatments that they had not tried was likely to meet with little success. We also thought human macrophage and dendritic cell lines may provide a key piece of information based on the success of the Wobus study with murine NoV. However, Dolin et al., and Schrieber et al., described pathology involving epithelial cells of the proximal small intestine in infected human volunteers (Dolin et al. 1975; Schreiber et al. 1973). Further evidence provided by the groups examining the role of the attachment of virions to cell-expressed blood group antigens led us to focus on human epithelial cell lines providing those characteristics.

## MODEL DEVELOPMENT USING IN VITRO 3-D CELL CULTURES

Our hypothesis is that the key to developing a representative *in vitro* pathogenesis model for NoV lies in the ability to grow and differentiate intestinal epithelia in a manner that approaches physiological functionality observed *in vivo*. Representative models of differentiated human intestinal epithelium can be established by growing the cells three-dimensionally (3-D) on collagen-I coated porous microcarrier beads in rotating wall vessel (RWV) bioreactors that model the physiological fluid shear environment in their respective organs (Nickerson et al. 2001; Nickerson et al. 2005; Nickerson et al. 2004). The design of the RWV bioreactor is based on the principle that organs and tissues function in a 3-D environment and that this spatial context is necessary for development of cultures that more realistically resemble *in vivo* tissues and organs (Abbott 2003).

Our starting point for developing an infectivity assay for human NoV was to use the 3-D INT-407 (ATCC CCL-6) small intestinal epithelial model. There were multiple factors that were considered for choosing this model. First, early human volunteer biopsy studies indicated that NoV infection targets the human small intestine (Dolin et al. 1975; Schreiber et al. 1973). Second, reports showing differentiation of INT-407 cells in 3-D in the RWV essentially produces a "co-culture" model of multiple intestinal cell types (enterocytes, goblet cells, and indications that M cell-like markers are also are also present) (Nickerson et al. 2001); this phenomena of multicellularity has been hypothesized as a factor likely needed for NoV infectivity (Duizer et al. 2004). And finally, extensive characterization of this model 3-D system (Nickerson et al. 2001; Nickerson et al. 2004) revealed that cell-surface antigens, which are thought to be important in the attachment of noroviruses to cells (e.g. Lewis$^{a}$), (Hale et al. 2000; Harrington et al. 2002; Harrington et al. 2004; Hutson et al. 2003; Lindesmith et al. 2003; Marionneau et al. 2002; White et al. 1996), are expressed under these growth conditions. Here, we present the methods and results of our first attempts to infect *in vitro* organoid 3-D intestinal models with human genogroup I and II human NoV.

# CHAPTER 2 MATERIALS AND METHODS

## DETAILED INSTRUCTIONS FOR PROPAGATING 3-D SMALL INTESTINAL CELLS

### Introduction

The *in vitro* cell culture method for noroviruses requires specialized cell culture apparatus that supports the growth and differentiation of cells that allow them to form tissue assemblies that mimic both the 3- dimensional architecture and physiological characteristics of the *in vivo* organs that are targeted by these viruses. While several valid methods exist for generating these differentiated cells (e.g. matrigel, trans-well inserts, etc. (Abbott 2003)), we chose the rotating wall vessel (RWV) bioreactor system reviewed by Nickerson and Ott (Nickerson et al. 2004) for all work reported here. The reasons for choosing this system were several. First, there is a growing body of literature that uses this system for studying pathogenesis of *Salmonella* and *E. coli* in both small and large intestinal models (Carvalho et al. 2005; Honer zu Bentrup et al. 2006; Nickerson et al. 2001), *Pseudomonas aeruginosa* in lung models (Carterson et al. 2005), cytomegalovirus in placental tissues (Lamarca et al. 2005), and Epstein-Barr virus in lymphoblastoid models (Long et al. 2001; Long et al. 1999). Second, unlike static 3-D systems the RWV bioreactor is a dynamic system that allows the cells to grow under conditions of normal physiological fluid shear. This may be a critical factor to achieve the level of cellular differentiation required for successful norovirus infection. Third, once the cell culture model has been built and characterized, the results are reproducible between batches of cells.

### Specialized Equipment Required

The physical equipment required for this work can be purchased from Synthecon (Houston, TX, www.synthecon.com). The company sells both disposable and autoclaveable bioreactor units. While more expensive, we recommend the reusable systems as these can accept both the reusable and disposable bioreactor vessels. The RCCS-1 unit comes with a base station, a power supply (connected from outside the $CO_2$ incubator to the base station via a flat ribbon cable) and either a reusable slow turning lateral vessel (STLV) or a high-aspect rotating vessel (HARV). The RCCS-4 unit is a 4-place base station with 1 power supply that comes equipped with either 4 STLVs or 4 HARVs. The initial capital expense for the RCCS-1 and RCCS-4 is approximately \$4,500 and \$14,000 USD, respectively. All cell culture work reported here, has been optimized using the ***reusable*** 55 mL STLV (Figure 2.1, left panel) and the ***disposable*** 50 mL HARV (Figure 2.1, right panel) bioreactor vessels, respectively. Disposable STLVs do not exist.

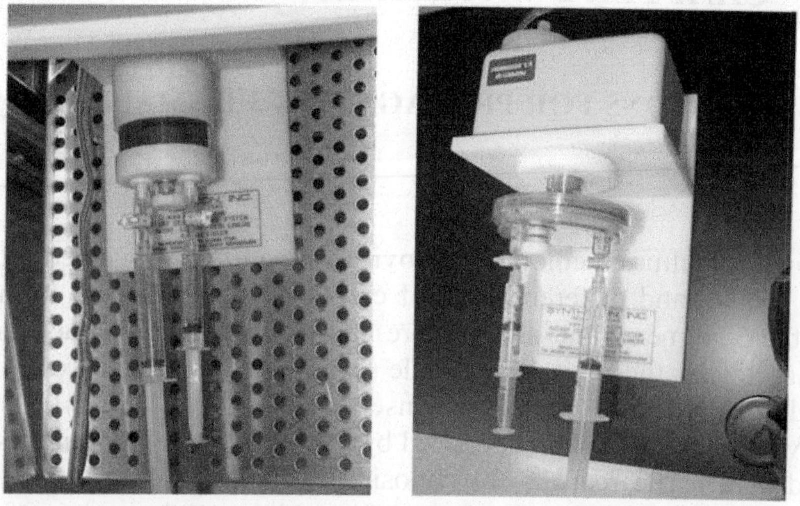

STLV HARV

**Figure 2.1 A RCCS-1 rotating wall vessel bioreactor unit illustrating the 55 mL reusable slow turning lateral vessel bioreactor (STLV) and disposable 50 mL high aspect rotating vessel (HARV), respectively. Power is provided via a ribbon cable that is connected to an external power supply.**

**Specialized Cell Culture Media Required**

GTSF-2 (HyClone, Logan, UT, USA, Part number SH3A099.01) is a specialized growth media that is used for generating these 3-D tissue assemblies (Lelkes et al. 1997). HyClone produces this in custom batches of 100 or 200L. It is imperative to purchase the dry powder formulation vs. the liquid form for best results. Because it is a custom formulation, the price structure is unusual: 50L costs approximately the same as a 200L order (ca. $4,536). Thus, we recommend buying the 200L quantity because the unit cost/L is about half of the 100L batch. Currently, Wyle laboratories, a NASA contractor laboratory, controls access to this media through HyClone, and permission must be obtained from this laboratory before HyClone will allow users to order the media. However, since the media was developed with US government funds, any investigator who is associated with a federal laboratory or a federally funded grant is supposed to have access to this media. In addition, if the media is purchased as 5L bottles, this permission may not be necessary.

HyClone's base media is a triple sugar (glucose, galactose, and fructose) 40% alpha minimal essential medium and 60% Leibovitz's L15 medium. To make 1L of the tissue culture medium from this base:

- Weigh out 18 g GTSF-2 (Leibovitz medium in powder form).
- Weigh out 2.25 g $NaHCO_3$.
- Pour this into a sterile 1L bottle with approximately 700 mL of Filter sterilized milliQ water. *Perform this step under aseptic conditions. Use the laminar flow hood for this step and all subsequent steps.*
- Shake vigorously to dissolve the powder.
- Add 100 ml heat inactivated fetal bovine serum (10% final concentration).

- Add 5 ml Insulin transferrin sodium selenite (ITS, final concentration in media is 0.5X) (Sigma part number I3146).
- Add 2 ml Fungizone (final concentration in media is 0.5 μg/mL, Invitrogen part number 15290 – 018).
- Fill to 1L.
- Sterile filter and aliquot into 50 mL sterile conical centrifuge tubes.
- Store protected from light at 4°C.
- Only warm up as much medium as needed for feeding the cells. (Reason to aliquot into 50 mL tubes).
- Add Pen/Strep (0.5 mL of a 100X solution, final concentration in media is 100U penicillin and 100 μg of streptomycin per mL, respectively, Invitrogen part number 15140-155) to these aliquots before feeding the cells. The pen/strep solution is not stable when it is repeatedly heated.

**Propagating the 3-D Cells**

The majority of this section is described in Straub et al. (Straub et al. 2007). As we are learning these techniques from the Nickerson Laboratory, we have made some modifications that make this procedure somewhat easier. These modifications are described in much greater detail below.

The human embryonic intestinal epithelial cell line INT-407 can be obtained from the American Type Culture Collection (CCL-6) and is initially grown as standard monolayers in GTSF-2 medium (Hyclone, Logan, UT), as described above, in T-75 flasks at 37°C in a 5% $CO_2$ environment in preparation for seeding into the rotating wall vessel bioreactor (RWV). Cells are trypsinized at 70% confluency and a total of 5 $x10^6$ cells are added to the RWV. Cells are assayed for viability by trypan blue dye exclusion. Cells are then introduced into the RWV (Synthecon, Inc, Houston, TX) containing 5 mg/mL porous Cytodex-3 microcarrier beads (collagen type-I-coated porous microspheres, average size 175 microns in diameter, Sigma, St. Louis, MO), resulting in a final ratio of approximately 10 cells/bead.

To make the entire process described above more straightforward, and to insure that the cells attach to the beads prior to introduction into the RWV bioreactor, we produce 250 mg batches of the Cytodex-3 beads in 5 mL of 1X phosphate buffered saline (PBS). To 2-24 well tissue culture plates we add 75μL of the beads, such that the beads form a monolayer on each well. Trypsinized cells are washed 2X with 1X Hanks Balanced Salt Solution and then brought to 24 mL final volume in GTSF-2. To each well, 0.5 mL of cell suspension is added to the wells, and then the plates are incubated overnight at 37°C. The next day, the cells are examined to obtain a better estimate of the actual ratio of cells per bead. If acceptable colonization is observed (e.g. 2 – 10 cells per bead), a wide bore 10 mL pipette is used to transfer the newly formed tissue aggregates to the STLV bioreactor through the main filling port.

Once the aggregates are transferred to the bioreactor the main filling port is closed and media is added through the two secondary ports. If an STLV bioreactor is used, attach two sterile stopcocks (Medex One-way Stopcock w/ Swivel Male Luer lock, Fisher Part Number NC9889364) to the secondary filling ports. The HARV has these stopcocks built into the unit. For either vessel, aseptically remove the plunger from a 10 mL and 5 mL syringe, and store these aseptically. Make sure the stopcocks are open. Then add GTSF-2 media until the bioreactor is completely full. Add enough media such that the 5 mL syringe has 4-5 mL of media when both sides are at equilibrium. Remove all bubbles in the bioreactor, and then tilt the bioreactor slightly so that the 5 mL syringe is almost full of media. Insert the plunger, and carefully depress the plunger so that the media begins filling the 10 mL syringe (depress to final volume of ca. 1 mL in the 5 mL syringe). Insert the plunger for the 10 mL syringe. Make sure that all bubbles are expelled from the bioreactor. Close the stopcock on the 10 ml syringe. Depress the 5 mL syringe just until there is noticeable resistance. With this plunger depressed, close the stopcock. Make a final check for bubbles and attach the bioreactor to the base, and turn on the unit and begin rotation.

To change the media in either the STLV or HARV, begin by adding 500 μl of Pen/Strep (100X solution) to a 50 ml aliquot of GTSF-2 media and warming in a 37°C water bath. Turn off the power to the bioreactor, and remove the bioreactor from the $CO_2$ incubator. Change all media in a sterile tissue culture hood. Use an ethanol-soaked KimWipe or equivalent to disinfect the outside of the unit and place the unit into the sterile laminar flow hood. With the unit in the sterile hood, and with both stopcocks closed, carefully pull up on the plunger of the 10 mL syringe, and then carefully unscrew the syringe from the stopcock. Dispose of this media into an autoclavable container. Repeat for the 5 mL syringe. Dispose of both syringes in a biohazard bag. Alternatively, they can be rinsed with Wescodyne or equivalent disinfectant and disposed of in the regular garbage. Use an ethanol soaked KimWipe or equivalent tissue to clean any media that may have leaked or spilled from the stopcock. Position the bioreactor over the autoclavable container with the secondary filling ports at approximately the 10 and 4 O'clock positions. Allow the aggregates to settle. Open the lower stopcock and allow the media to start draining. Then open the upper stopcock to keep this process going. Drain enough of the media such that the aggregates are still covered with some media. Using new and sterile 10 and 5 mL syringes, repeat the process described above for adding sterile GTSF-2 media to the bioreactor.

Cells are cultured in the RWV bioreactors with the rotation speed initially set at 18 RPM to maintain the cell aggregates in suspension during the entire culture duration. In general, this rotation speed will suffice throughout the incubation period. In very rare cases, highly efficient growth may result in settling of the tissue aggregates. If that is the case, the rotation speed needs to be increased, sometimes as high as 24-28 RPM. Fresh medium is replenished by 90% of the total vessel volume each 24 hours to 5 days, depending upon the growth and metabolic rate of the cultures. In general, the medium is changed once at day 5, every 48 hrs during days 5-14, and then every day until Day 28. On Day 28, use a wide bore 10 mL pipette and transfer the 3-D aggregates from the STLV bioreactor to the disposable HARV bioreactor, and fill this bioreactor with media (instructions above). Continue changing the media every day in the HARV until day 35.

With the exception of the 4th infectivity trial described in Chapter 3 of this report, 3-D cells are harvested no later than 35 days after seeding into the RWV. For the $4^{th}$ infectivity trial, aggregates were harvested starting on day 29 and continuing to day 35. Using a 10 ml wide-bore pipette, mature 3D aggregates are first transferred to a 50 mL conical centrifuge tube, and washed 3X with serum-free GTSF-2 to serum starve the cells. Then, using another wide bore 10 mL pipette, they are plated into 24-well plates (approximately 400,000 cells / well) and infected with varying dilutions of norovirus on the same day. Thus, the usable timeframe for the 3-D aggregates can be as soon as Day 28 (see next section to make that determination), but no later than Day 35.

## Verification of Differentiation – Immunohistochemical Staining

Prior to each infectivity assay, immunohistochemical staining is performed on aliquots of the 3-D INT 407 cells to ensure differentiation. ***It is imperative that these tests provide acceptable results prior to, or in conjunction with each norovirus infectivity assay. If these tests fail, it is likely that the infectivity assay will fail as well. Acceptable results are minimally defined here as correct localization of the tight junctional markers (cell surface expression (correct) vs. diffuse expression in the cytoplasm (incorrect)). In addition evidence of two cell types: the enterocytes and goblet cells, determined via PAS staining, and evidence of well-formed apical microvilli provide additional evidence of correct differentiation.*** Paraformaldehyde fixed (4% PFA in 1X PBS for 30 min at room temperature) aliquots of the mature tissue aggregates are stained with antibodies specific for these tight junction markers: ZO-1, Occludin, Claudin-1, and E-cadherin (Zymed Laboratories (Invitrogen), South San Francisco, CA) and imaged using confocal laser scanning microscopy (Zeiss Axioplan II microscope, Carl Zeiss, Thornwood, NY). Correct localization of these markers at cell membranes is highly indicative of differentiated cells. Previous characterization of the 3-D INT407 model also included collagen type-II, fibronectin, sialyl Lewis A antigen, villin and PAS staining to reveal mucin production, and these tests may be useful for laboratories adopting these methods for the first time (Nickerson et al. 2001). In addition, it is also good QC practice to either perform scanning electron microscopy or thin section light microscopy on the aggregates to insure that the model has produced numerous and well-formed apical microvilli. Refer to Nickerson et al. (Nickerson et al. 2001) to show results of well differentiated 3-D INT-407 cells vs. monolayers to compare results.

The procedure for antibody staining for the cell surface markers is as follows:

1. Fix the cells in 500µl of fixative (see below) for 30 min at room temperature in a light-free area.
2. Wash off the fixative with Delbecco's PBS (DPBS containing 1mM $Ca^2$+ and 1mM $Mg^2$+), Wash three times. Make sure that the PBS is applied onto the side of the wells as to not dislodge the cells when performing these washes.
3. Incubate sample with ~ 500µl 50mM $NH_4Cl$ in DPBS for 10-20 min at room temperature, to reduce background fluorescence.
4. Wash cells 3X with DPBS
5. Permeabilize the cells with ~ 500µl of 0.1% Triton-X 100 (Tx-100 in DPBS) for ~5-8 min at room temperature (add 5 µl of 10% Triton X-100 in 495µl total volume of DPBS with $Ca^{2+}/Mg^{2+}$, final concentration of Tx-100 = 0.1% - or just add ~ 500µl 0.1% TX-100).
6. **Optional: It is possible to combine steps 3 and 5 together, and then wash 3X with DBPS (step 4).**
7. Wash cells 3X with DPBS
8. Block by adding ~500µl of 8% heat inactivated (HI) goat serum in DPBS containing 0.05% TX-100 for 60 or more min at room temperature. This is to lower background staining by blocking unspecific binding of the fluorophores.
9. **Optional: Step 8: Can use 5% BSA in DPBS and incubate at 4°C overnight as an alternative to HI goat serum.**

10. Optional: Add 2μl of the Phallotoxin (e.g. Alexa Fluor 568 Phalloidin from Invitrogen/Molecular Probes. Quantity is 300 Units, part number A12380) to 200 μl of 0.8% HI Goat serum (with 0.05% Tx-100) plus cells. Incubate for 40 min at room temperature in a light-free area. The phalloidin intercalates with the F-Actin polymers allowing visualization of the cytoskeleton. This step is not necessary for viewing tight junctional markers.
11. If using antibodies: Add primary antibody at the correct concentration (in 0.8% Goat serum plus Tx-100) to the cells. Incubate for ~ 1 hr at room temperature.
12. **Optional substitution: Use 1% BSA in DPBS and incubate the primary antibody overnight at 4°C.**
13. Wash 4X with DPBS.
14. Add secondary antibody (if not using coupled Abs) at the correct concentration (in 0.8% Goat serum plus Tx-100) to the cells. If needed, include DAPI counter stain (add 1-10 μM DAPI to antibody solution). DAPI stains the nucleus by interacting with double stranded DNA. (DAPI Stock solution = 10 mM in dimethyl formamide). Incubate for about 1 hr at room temperature (or more if needed).
15. **Optional substitution for goat serum: Use 1% BSA in DPBS and incubate overnight at 4°C.**
16. Wash 3X with DPBS for 5 min each wash.
17. Remove the extra fluid from around the section and cells and add 100 μl Prolong-Antifade (Molecular Probes) per 50μl of 'beads'. To make the Prolong Mounting Medium, warm 1 ml of component B (glycerol) at 50°C for 30min (in an Eppindorf microcentrifuge tube). Add B to component A (antioxidant) and vortex thoroughly. Spin for 10 min at 13,000 rpm to remove air bubbles (may have to use a P20 tip). Keep cool and dark, freeze at -20°C after use.
18. With a wide bore yellow tip (just cut off the tip), place cells in mounting medium on the slide (use pre-cleaned slides.). Pop air bubbles with needle, place coverslip (use # 1.5) on slide in an angle so as to avoid any air bubbles underneath. Make sure that there is mounting medium underneath the whole coverglass.
19. Let dry in the dark overnight at room temperature.
20. When dry, apply CLEAR nail polish to the edges of the cover slip to preserve, and seal off, store in the dark at 4°C.
21. Image as soon as possible using a fluorescence or confocal laser-scanning microscope because the antifade tends to increase in background fluorescence after several days.

**Fixative requirements additional notes:**

Formaldehyde fixative is 2% for monolayers, and 4% for 3-D cells, and a total exposure of 30 min at room temperature. Make sure you use either Paraformaldehyde (PFA) or EM grade formaldehyde; it needs to be methanol free. The Nickerson laboratory now uses this PFA fixative for all of their antibody work. The Working Formaldehyde Fixative is 2% (or 4%) formaldehyde (EM grade) made up in 1X D-PBS (pH 7.4). This is to be made fresh every time.

## METHODS DESCRIBED IN EMERGING INFECTIOUS DISEASES (STRAUB ET AL. 2007)

### Preparation and Characterization of Virus Stocks

Stool samples were obtained from acute gastroenteritis outbreaks from cruise ships (identified as samples 149 (GII) and 155 (GI)) and a nursing home (identified as flag2, a GII strain). Approximately 1 gram of stool was suspended in 0.01 M PBS to obtain a 10-20% stool suspension (ca. 5 – 10 mL). The suspension was vortexed for 60 seconds, centrifuged at 1,000 x g and filtered through a 0.22 micron filter to remove bacterial contamination. Virus samples were stored at -80°C for future assays.

The presence of norovirus in the purified samples was confirmed by reverse transcription-PCR analysis (Vinje et al. 2004) and genotyped by sequencing and BLAST to GenBank (Table 2.1). Stool extracts were screened for other enteric viruses by 3 passages on Buffalo Green Monkey cells and Caco-2 cells grown as conventional monolayers. Stool extracts were also tested for enterovirus by Reverse Transcription-PCR as described previously (Fout et al. 2003).

Viruses, diluted 1:5 to 1:1000 in serum-free GTSF-2 media, were applied as 0.1 mL aliquots per well of a 24-well plate across a minimum of 3 wells per time point for any of the infection trials described in Table 2.2. Viruses were introduced to serum-starved cells by gentle mixing of the aggregates with the viral suspension. The infected aggregates were incubated for 1 hr at 37°C in a 5% $CO_2$ incubator before being overlaid with 1mL of fresh GTSF-2 (containing serum) media.

**Table 2.1**
**Genetic characterization of the RNA dependent RNA polymerase sequence of the norovirus strains used in this study (Straub et al. 2007).**

| Sample ID | Date Collected | Setting | Strain in GenBank with closest sequence similarity (%) | Genogroup |
|---|---|---|---|---|
| 149 | 14 April 2004 | Cruise ship | AJ487474 NLV/Castell/2001/Sp (97%) | II |
| 155 | 21 June 2004 | Cruise ship | DQ157140 Hu/Offenburg1155/2004/D (100%) | I |
| flag2 | 14 December 2004 | Nursing home | AJ626578 NV/GII/Stockholm/IV1138/2003/SE (97%) | II |

**Table 2.2**
**Summary of Methods from the 4 infectivity trials described in this report (Straub et al. 2007)**

| Infectivity trial and date | Virus stocks used and comments regarding the trial. | Time points assayed | Assays performed |
|---|---|---|---|
| 1st March 2005 | Combined equal volumes of strains 149, 155, and flag2 (P0)*. Effective dilutions of $10^{-1}$ to $10^{-3}$ were assayed. Supernatants† from all infected wells, dilutions of viral stock, and time points combined (ca 15 mL final volume). | 1 hr to 72 hrs, media changed in all wells at 24 hrs post infection (pi). | CPE, reverse transcription PCR (RT-PCR), thin section light microscopy and ultra-thin section TEM. |
| 2nd June 2005 | Supernatant cocktail from first infectivity trial (P1), strains 149 and flag2 tested alone (P0 stool samples). Supernatants from each time point were harvested for subsequent infectivity trials (ca. 3mL/time point) | Same as first infectivity trial, media changed every 24 hours. | Same as first infectivity trial except that CPE was documented photographically. |
| 3rd August 2005 | Supernatant from combined stock (P2), 149, and flag2 (P1), stool sample flag2 (P0): Controls generated by ultrafiltration (10,000 MWCO). | Same as 2nd infectivity trial. | Same as second infectivity trial, first attempt with Fluorescence in situ hybridization (FISH). |
| 4th December 2005 | Infectivity followed through 5 passages in cell culture using strains 155 and flag2 (P0 – P5). Effective dilution of viruses at P5 = 1:$10^6$, if replication was not occurring. Passages were spaced every 24 hrs. | Infected aggregates were processed at 24 hrs pi, and the viruses were used for subsequent passage. | CPE, RT-PCR, Molecular beacon FISH. PCR products from P3 of both strains were cloned and sequenced. Compared sequences with sequenced PCR reactions from original stools. |

* Passage number definitions referred to in table and text: P0: viruses from a stool sample and used to infect a cell culture for the 1st time. P1: viruses harvested from P0 cell cultures and used to infect cell culture a 2nd time. P2: viruses harvested from P1 and used to infect cell culture a 3rd time. P3: viruses harvested from P2 and used to infect cell culture a 4th time. P4: viruses harvested from P3 and used to infect cell culture a 5th time. P5: viruses harvested from P4 and used to infect cell culture a 6th time.

† Supernatants are defined as the liquid fraction and any cells that had left the intact tissue aggregates.

## Microscopic Analysis (Straub et al. 2007)

Cellular pathology of 3-D tissue aggregates for the second and third infection trial was documented using an Olympus DP70 CCD camera and inverted microscope system (Nikon Eclipse TE300) at each time point assayed. Sub-cellular pathology was assessed using light and transmission electron microscopy. These samples (approximately 1 x $10^6$ cells per well) were fixed in 3.5% glutaraldehyde, 0.5 % paraformaldehyde, in PBS and processed by washing cells three times with 0.1 M sodium cacodylate buffer (Electron Microscopy Sciences) prior to incubation in 1% osmium tetroxide diluted in 0.1 M sodium cacodylate buffer for one hr at room temperature. Cells were washed with buffer and dehydrated using a graded series of ethanol rinses (33, 50, 75, 90, and 3 x 100% Ethanol (EtOH)). Samples were then embedded in hard grade LR WhiteTM Resin (London Resin Co., Berkshire, England) at 60°C for 24 hrs. Block faces were cut into the samples using a Leica EM Trim (Wetzlar, Germany). Thin and ultrathin sections (60 nm and 30 nm) were cut using an Ultra 45° Diatome (Biel, Switzerland) diamond knife on a Leica Ultracut UCT ultramicrotome. Thin sections were affixed on microscope slides and stained with toludine blue before viewing on a Nikon Optiphot-2 (Kanagawa, Japan) light microscope. Ultra-thin sections were affixed on copper mesh grids and were stained and contrasted with uranyl acetate and lead citrate for seven minutes each before viewing on a JEOL-2010 (Tokyo, Japan) transmission electron microscope at 106 kV.

## Viral RNA extraction and RT-PCR (Straub et al. 2007)

RNA from tissue samples was extracted either using an RNEasy or ViralAmp RNA extraction kit (Qiagen, Valencia, CA). Reverse transcription PCR was performed using the OneStep RT-PCR kit (Qiagen) following the manufacturers' instructions. Primer sequences for RT- PCR and semi-nested PCR to amplify the RNA dependent RNA polymerase gene (Region A) are listed in Vinje et al. (Vinje et al. 2004), with the exception of the MP290 primer for semi-nested PCR (MP290 5'-GAYTACTCYCSITGGGAYTC-3'). Viral RNA was reverse transcribed for 60 min at 42°C, 15 min at 95°C to inactivate the RT enzyme, and activate Taq, followed by a 3-step PCR for 40 cycles (30 sec at 94°C, 30 sec at 50°C, and 30 sec at 72°C).

## Sequencing (Straub et al. 2007)

PCR products amplified from cell cultures (P3 passage) of the 4th infection trial were sequenced after cloning into the PGEM-T Easy Vector System (Promega, Madison, WI) per manufacturer's instructions. Sequences have been deposited in GenBank (Accession numbers are DQ531707 for outbreak sample 155 and DQ531708 for outbreak sample flag2, respectively).

## Fluorescence In Situ Hybridization (FISH)(Straub et al. 2007)

The molecular beacon FISH assay employed during the 4th infection trial used the following probe sequences which were modified reverse PCR primer sequences for genogroup I and II viruses, respectively (Kageyama et al. 2003). For genogroup 1, the modified probe sequence is 5'- TAMRA - <u>CAGGC</u>CCTTAGACGCCATCATCATT<u>GCCTG</u>-DABCYL - 3', and for genogroup 2, the modified probe sequence is 5' –TAMRA - <u>CTCGGT</u>CGACGCCATCTTCATTCAC<u>ACCGAG</u>-DABCYL – 3' (underlined sequences for each probe indicate the self-complimentary regions). Cells in the tissue aggregate were fixed in 4% paraformaldehyde for 30 min, and then washed 3 times in 1X Dulbecco's phosphate buffered saline (DPBS, Sigma Chemical Co., St. Louis, MO. USA). The weight of the aggregates allowed these to settle to the bottom of the microfuge tube by normal gravity. Tissues were permeabilized with 0.1% Triton X-100 in 1X DPBS for 15 min at room temperature, and then washed 3 times with 1X DPBS. Molecular beacon (either genogroup 1 or genogroup 2) was suspended to a final concentration of 1 μM in 1X DPBS. The molecular beacon was incubated with the tissues in a water bath for 1 hr at 37°C. The aggregates were then washed 3 times with 1X DPBS, and transferred to 12-well tissue culture plates. Cell aggregates were imaged using a Leica confocal laser-scanning microscope using a 63X water immersion objective. Captured images were digitally stacked to create three-dimensional images (VOLOCITY, Improvision Inc., Lexington, MA).

## METHODS NOT REPORTED IN STRAUB ET AL. (2007)

### Development of standards for quantitative real-time PCR protocols

Quantitative reverse transcription real-time PCR was performed to determine the degree of productive viral RNA replication in cell culture. Our protocols essentially followed those described in Kageyama et al. (Kageyama et al. 2003), with the following exceptions. Plasmid standards were designed to target the highly conserved ORF1-ORF2 junction. Genogroup I standards were developed from sample 155 described above. PCR primers were designed to produce a fragment that was slightly longer than the segment amplified by real-time PCR. The primer and cloned product sequences for Genogroup I are:

- F: 5'-AGGCCATGTTCCGCTGGAT-3'.
- R: 5'-GCGTCCTTAGACGCCATCATCATT-3'.
- The sequence of the 101 nt amplicon that was cloned to make the plasmid standard is: 5' - <u>AGGCCATGTTCCGCTGGAT</u>GCGCTTCCATGACCTCGGATTGTGGACAGGAGA TCGCAATCTCCCGCCCGAATTCGTA<u>AATGATGATGGCGTCTAAGGGCGC</u>-3' (Underlined sequences indicate the position of the flanking primers.)

For genogroup II, plasmid standards were produced from a cruise ship isolate (identified as sample VS122) sequenced as a GII.4 strain (Hunter: per sequence of the capsid).

- F: 5'-TGCCCAGACAAGAGCCAATGTT-3'.
- R: 5'-GTCACTCGACGCCATCTTCATT-3'
- The sequence of the 111 nt amplicon that was cloned to make the plasmid standard is: 5'- <u>TGCCCAGACAAGAGCCAATGTT</u>CAGATGGATGAGATTCTCAGATCTGAGCAC GTGGGAGGGCGATCGCAATCTGGCTCCCAGTTTTGTAATG<u>AAGATGGCGTCG AGTGAC</u> -3' (Underlined sequences indicate the position of the flanking primers.)

In order to make the plasmid standards containing either the genogroup I or genogroup II sequence of interest, viral RNA was extracted from the stool samples as previously described, and subject to reverse transcription using Invitrogen's Superscript III protocol. Briefly, 1 µL of random primers (100 ng/µL), 1 µL of dNTPs (10 mM), and 11 µL of viral RNA were mixed and incubated for 5 min at 65°C followed by chilling on ice for 1 min. Then, 4 µL of 5X first strand buffer, 1 µL of 0.1M DTT, 1µL of RNase Out (40U/µL) and 1 µL of Superscript III (200 U/µL) was added in order to bring the final reaction volume to 20 µL. The first strand reaction conditions consisted of 5 min at 25°C, followed by 60 min at 50°C, 15 min at 70°C, and terminating with an indefinite 4°C step. First strand cDNA templates for both genogroup I and genogroup II viruses were then subjected to PCR.

PCR amplification was performed in 25 µL reaction volumes and consisted of 2 µL of first strand cDNA template, 5 pmol of forward and reverse primers for each genogroup, 200 µM dNTP, 1.5 mM $Mg^{2+}$, 1X PCR buffer, 2.5 U HotStar Taq Plus (Qiagen), and sterile water to volume. Thermal cycling protocols were 95°C for 5 min to activate the Taq Polymerase, then 40 cycles of denaturation at 95°C for 15 s, primer annealing at 50°C for 30 s, primer extension at 72°C for 45 s, and then a final extension period of 10 min at 72°C. Reaction products were analyzed using either agarose gel or capillary electrophoresis (Agilent 2100 bioanalyzer) to determine the correct size of the products. PCR products were purified using Qiagen's PCR purification kit. Products were cloned and sequenced as previously described.

Verification of the insert was performed by restriction enzyme digestion followed by capillary electrophoresis and PCR. Plasmid DNA concentrations were measured using a Nanodrop spectrophotometer. To a 20 µL total reaction volume, no more than 0.5 µg of plasmid DNA was added to 2 µL of EcoR1 buffer, 1µL (1 U) of EcoR1 restriction enzyme, and DNAase free water to volume. The reaction mixture was incubated at 37°C for 3 hrs, the reaction was stopped by heating to 65°C for 20 min, and finally chilled to 4°C prior to analysis. Capillary electrophoresis and PCR confirmed both the correct size and sequence of the insert, respectively. From this data plasmid standards were generated with verified culture clones. The following calculation example shows how ng of plasmid DNA translate to copy number:

- Insert size = 101 bp (Genogroup I)
- Plasmid + insert = 3101 bp
- Because plasmids are double stranded = 6202nt (total nucleotide bases)
- Assume that the average molecular weight (MW) of a nucleotide base is 330 daltons
- 1 mole of plasmid = 6202 bp x 330 = $2.04 \times 10^{6}$ g/mole or $2.04 \times 10^{15}$ng/mole
- 1 mole = $6.02 \times 10^{23}$ molecules
- We want $1 \times 10^{10}$ molecules (copies of plasmid)
- = ($2.04 \times 10^{15}$ng x $1.0 \times 10^{10}$ molecules)/($6.02 \times 10^{23}$ molecules)
- Thus, 34 ng of plasmid = $1.0 \times 10^{10}$ copies.

**Infectivity Trial Sample Processing**

Viral RNA from stool or cell culture was extracted using Qiagen's ViralAmp RNA kit per manufacturer's protocol with minor modifications discussed here. For stool or cell culture reverse transcription real-time PCR was performed by extracting 140 µL of stool, cell culture supernate or cell aggregates, separately. In the case of the cell aggregates, these were carefully washed 2X in 1 X PBS, and the supernate from these washes was discarded. The tissue aggregates were suspended in an appropriate volume of lysis buffer from the ViralAmp RNA kit. For all extractions the mixture of sample and lysis buffer was first passed through a

QIAshredder. This was helpful for the cell aggregates to remove the collagen beads and dead cell mass. After subjecting to the QIAshredder, viral RNA proceeded as per manufacturer protocol.

**Reverse Transcription Real-Time PCR**

First strand cDNA synthesis was performed as described above, and then real-time PCR was performed using an ABI 7500 real-time PCR instrument following Kageyama et al.'s (2003) protocol with a few minor modifications. We used 2 µl of cDNA in a final reaction volume of 20 µl. Primer and probe concentrations were as described by Kageyama et al. (2003). Titration of stool samples in GTSF-2 media (stool to 1:100 dilution of the stool) demonstrated that the Qiagen ViralAmp RNA kit removed potential inhibitors (e.g. approximate 10-fold decrease in copy number at each dilution). Calculation of total copies applied to cells was compared to calculations for total copies in cell culture to provide a direct comparison of results, and facilitate in quickly determining the level of viral production in cell culture.

**Biosignature Discovery (Data Sharing from Other Projects)**

DISCLAIMER: We are describing two methods for examining host metabolic volatile organic carbon and putative secreted host cell peptides in response to infection by human noroviruses. The details of these methods and the data are being shared in this report as a courtesy and are presented "as is" (e.g. unedited information provided by our collaborators). However their inclusion or exclusion would not change the overall conclusions of the work reported here.

**Preparation of cell culture supernatants for biosignature discovery**

All cell culture supernatants were passed through a 30,000 kilodalton molecular weight cutoff ultrafilter (YM-30, Millipore, Billerica, MA, USA) centrifuged at 12,000 x g for 12 min to remove the virus and higher molecular weight sera proteins. The sample was held at -80°C until thawed for further analysis.

**Gas chromatography x gas chromatography time-of-flight mass spectrometry for host secreted metabolic volatile organic compounds**

A Leco Pegasus 4 dimensional GC X GC -TOFMS system (LECO Corp., St. Joseph, MI, USA) equipped with a Gerstel mulitpurpose sampler (MPS2) for automated SPME analysis (GERSTEL Inc., Baltimore, MD, USA) was used for all sample analyses. A 10 mL headspace vial containing a SPME fiber was pre-cleaned and baked at 150°C for at least two days prior to adding 1 mL of cell culture supernatant. Samples were placed into a liquid-cooled sample tray equilibrated to 40°C on the day of analysis. Volatiles from the cell culture supernatant were captured on the SPME fiber. The fiber was then submitted for GCxGC TOFMS. The separation utilized ultra-high purity helium as the carrier gas, set at 1 mL/min, constant flow. Additional experimental details are listed in Table 2.3.

**Table 2.3**
**Experimental conditions for host-cell metabolic volatile organic compounds**

| | Settings | Manufacturer |
|---|---|---|
| GC column 1 | 30m x 250um x .025 um df<br>Initial: 35°C hold for 10 minutes,<br>5°C/min to 250°C,<br>250°C hold for 5 minutes | SolGel-Wax, (SGE, Inc. Austin, TX, USA) |
| GC column 2 | 2m x 100um x .4um df<br>+10°C offset from column 1 | 007-1701, (QUADREX CORP. Woodbridge, CT, USA) |
| Modulator | +25°C offset, 5 sec modulation period | |
| SPME sampling | Fiber: 2 cm stableflex, 50/30um Divinylbenzene/Carboxen/Polydimethylsiloxane (DVB/CARBOXEN-PDMS),<br>Extraction: 40°C for 68 minutes | (Supelco, Bellefonte, PA, USA) |
| GC injection/desorption | 5 min split less, 60 sec purge time @ 250°C | |
| MS | 70 eV, m/z 35-400 @ 100 Hz | |

HIMAS™detector (Bruker Daltonics). Protein samples were spotted on a ground steel 384 multi-target plate using 0.5 µl of sample followed immediately by 0.5 µl of matrix solution. Sinipinic acid (SA) matrix was used as a 10mg/ml solution in 70% acetonitrile/ 0.03% TFA. An external mass calibration was performed using protein standard mix II (Bovine myoglobin, cytochrome c, ubiquitin and insulin) spotted on adjacent locations. The spotting and analysis order was randomized by using the RAND function within Microsoft Excel, prior to loading the plate into the instrument.

Data was collected in linear mode from 2,000 to 30,000 m/z using a pulsed ion extraction delay time of 120 nano sec. The resulting mass spectra consisted of 500 laser shots summed together. Data was processed using available functions in the vendor software including a baseline subtraction and smoothing function (Savitzky-Golay with 0.2 m/z window) followed by peak picking using the centroid function for peaks with s/n greater than 5.

***Mass spectral data comparison*** Mass spectra were compared using Clinprotools 2.0 (Bruker Daltonics). This data analysis tool allows for spectral recalibration and alignment prior to comparison of treatment groups. Spectra from a similar treatment group (same treatment, time point and cleanup method) were analyzed together as a class. Realignment and recalibration was performed using the following settings within the software. Spectral preparation utilized the Convex Hull baseline function over a 2000-50,000 m/z range. Peak picking utilized a noise reduction threshold ignoring peaks s/n<2 and picking peaks with s/n > 5. The recalibration used a minimum resolution of 100 with a 0.2% maximum peak shift requiring 30% match to the calibrant peaks. Sample classes were compared using all peaks for model generation and prioritization of peak areas for determining indices (peaks) providing greatest discriminating power.

# CHAPTER 3: RESULTS FROM THE FOUR INFECTIVITY TRIALS

## FIRST INFECTION TRIAL MARCH 2005

Based on a hypothesis that a combination of diverse strains might be necessary to initiate infection and replication *in vitro*, the first infection trial combined three antigenically diverse norovirus stool samples. Samples 149 and Flag2 were sequenced as genogroup II isolates, and sample 155 was sequenced as a genogroup I isolate (Table 2.1 of this report). Dilutions of this combined stock (effective dilutions of 1:10 to 1:1000 from the original combined stool stock) were assayed to estimate a preliminary dose response/TCID50.

Within 24 hours post infection (pi) gross pathology of the infected vs. the uninfected 3-D small intestinal tissue aggregates, as observed by phase-contrast microscopy, indicated the possibility of norovirus infection for all viral dilutions. Cytopathic effect (CPE) was first observed in tissue aggregates receiving the greatest concentration of virus, followed by CPE developing in aggregates receiving the least concentration of virus. Thin section light microscopy provided more evidence of viral infection of the infected tissue aggregates (Figure 3.1). Consistent with the findings of Dolin et al., and Schreiber et al., infected tissue exhibits shortening of the microvilli and vacuolization vs. healthy tissue (Dolin et al. 1975; Schreiber et al. 1973) (Figure 3.1, Panels A and B for uninfected vs. infected tissue aggregates, respectively).

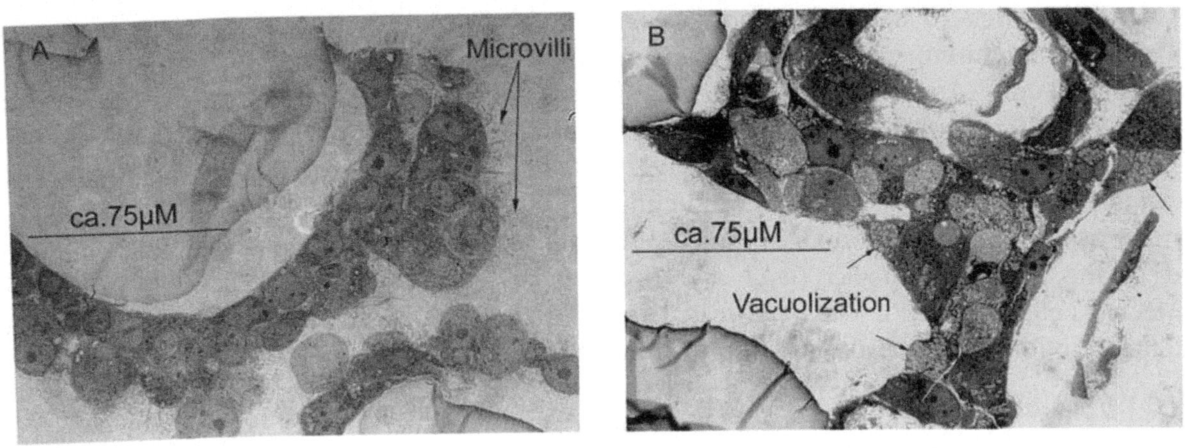


**Figure 3.1 Light microscopy of uninfected (A) vs. NoV infected 3-D INT 407 cells (B).**

Transmission electron microscopy revealed the presence of uniform ca. 29 nm diameter particles (consistent for the size of norovirus particles) invading the microvilli within 1 hr pi (Figure 3.2 Panel A) and accumulation of these particles within the tissue aggregates within 24 - 66 hrs pi (Figure 3.2 Panels B and C, respectively).

Reverse transcription - PCR (RT – PCR) was performed on all viral dilutions assayed in cell culture to determine if there was a relative increase in viral RNA. Table 3.1 illustrates the results of these experiments. At 1 and 24 hrs pi, tissue aggregates receiving the 1:10 and 1:100 dilution of the working virus stock were positive by RT - PCR but the aggregates receiving the 1:1,000 dilution of the virus stock were negative. However, by 66 hrs and 72 hrs pi the 1:1000 virus dilution tested positive by RT – PCR, indicating a potential increase in viral RNA. Uninfected tissue aggregates were negative by RT – PCR for all time points tested.

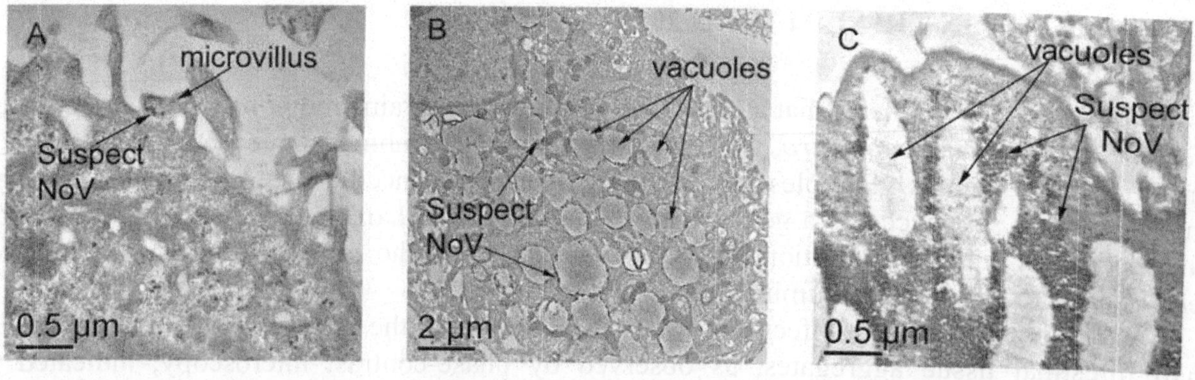


**Figure 3.2 Progression of NoV infection in cells. (A) 1 hr pi showing NoV involvement of the microvilli. (B) Suspect NoV particles at 24 hrs pi. Note the rearrangement of internal membrane structures. (C) 66 hrs pi showing significant involvement of the cells.**

## SECOND INFECTION TRIAL JUNE 2005

The second infection trial investigated whether viruses harvested from the first infection trial could be used to cause infection of naïve (uninfected) tissue aggregates. It was also important to determine whether individual strains, when tested separately, could infect these differentiated cells. As with the first infection trial, evidence of viral CPE was manifested within 24 – 48 hrs. Figure 3.3 displays the results of uninfected (Figure 3.3 Panel A) vs. infected (Figure 3.3 Panels B, C, and D) tissue aggregates. Viruses used to infect tissue aggregates shown in Panel B originated from the cell culture lysates of the first infection trial. Viruses used to infect tissue aggregates shown in Panels C and D originated from two of the outbreak strains used in the combined stock: flag2 and Sample 149, respectively. When compared with uninfected cells, viral CPE was manifested by cells sloughing off the collagen beads as a mat, with the individual cells becoming highly elongated or distorted.

**Table 3.1**
**Relative increases in viral RNA in infected tissue aggregates as measured by end-point dilution PCR**

| | Hours post infection | | | |
|---|---|---|---|---|
| Effective dilution of working viral stock applied to cells | 1 hr | 24 hr | 66 hr | 72 hr |
| 1:10 | + | + | + | ND |
| 1:100 | + | + | + | ND |
| 1:1000 | - | - | + | + |
| Negative control | - | - | - | - |

Transmission electron microscopy of the uninfected vs. infected aggregates is shown in Figure 3.4. All panels in Figure 3.4 were taken from tissue aggregates at 48 hours post infection. The uninfected control (Panel A) displays an intact Golgi body whereas in infected samples (Panels B, C, and D), we failed to find intact Golgi bodies. This finding is important in light of two studies. One by Fernandez-Vega et al. (Fernandez-Vega et al. 2004) that showed that non-structural proteins from NoV can cause dissociation of the Golgi. The other study conducted by Wobus et al. (Wobus et al. 2004) demonstrated similar internal membrane rearrangement in their TEM photographs when murine noroviruses infect macrophages. For infected – but not uninfected - cells, we always observed numerous uniformly sized 27 - 29 nm particles, which would be consistent for human norovirus (Kapikian et al. 1972). As orthogonal evidence that the particles we observed in the TEM micrographs were noroviruses, RT-PCR was positive for all infected samples shown in Panels B, C, and D. *In fact, end-point dilution PCR indicated that both the P1 infection with the combined stock from the 1$^{st}$ infection trial (Panel B) and flag2 (Panel C) sample increased in viral RNA titer over the course of the experiment.* In contrast, even after semi-nested PCR, the uninfected tissue aggregates were PCR-negative.

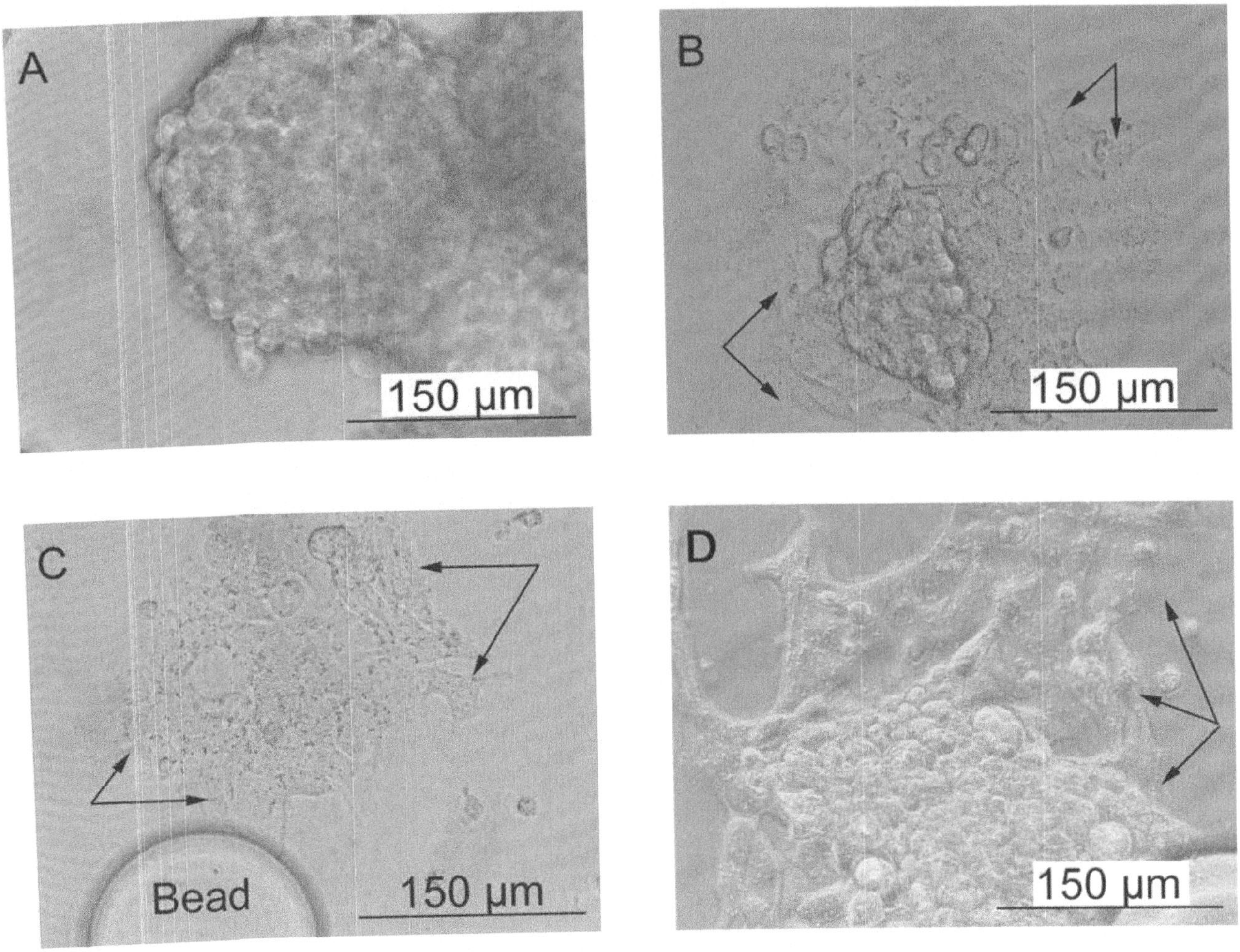


**Figure 3.3 Demonstration of cytopathic effect in infected tissue aggregates during the second infection trial. A) Uninfected aggregate, 24 h into the experiment. B) Cells infected with lysate from the first infection trial (P1) at 24 h pi. C) Stool sample flag2 at 24 h pi (P0). D) Stool sample 149 at 48 h pi (P0). Arrows indicate cells exhibiting unusual pathology (Straub et al. 2007).**

**THIRD INFECTION TRIAL AUGUST 2005.**

The third infection trial sought to address several issues we observed from the first two infection trials. The first question concerned the nature of CPE. Was the cytopathic effect due to potential toxins in these samples or most likely due to viral infection? The second question we asked was whether we could demonstrate a second passage of at least one of the strains to naïve cell cultures. Finally, the third question concerned the combined norovirus stock harvested from the second passage in cell culture and whether it could be used to cause infection in a 3rd (P2)

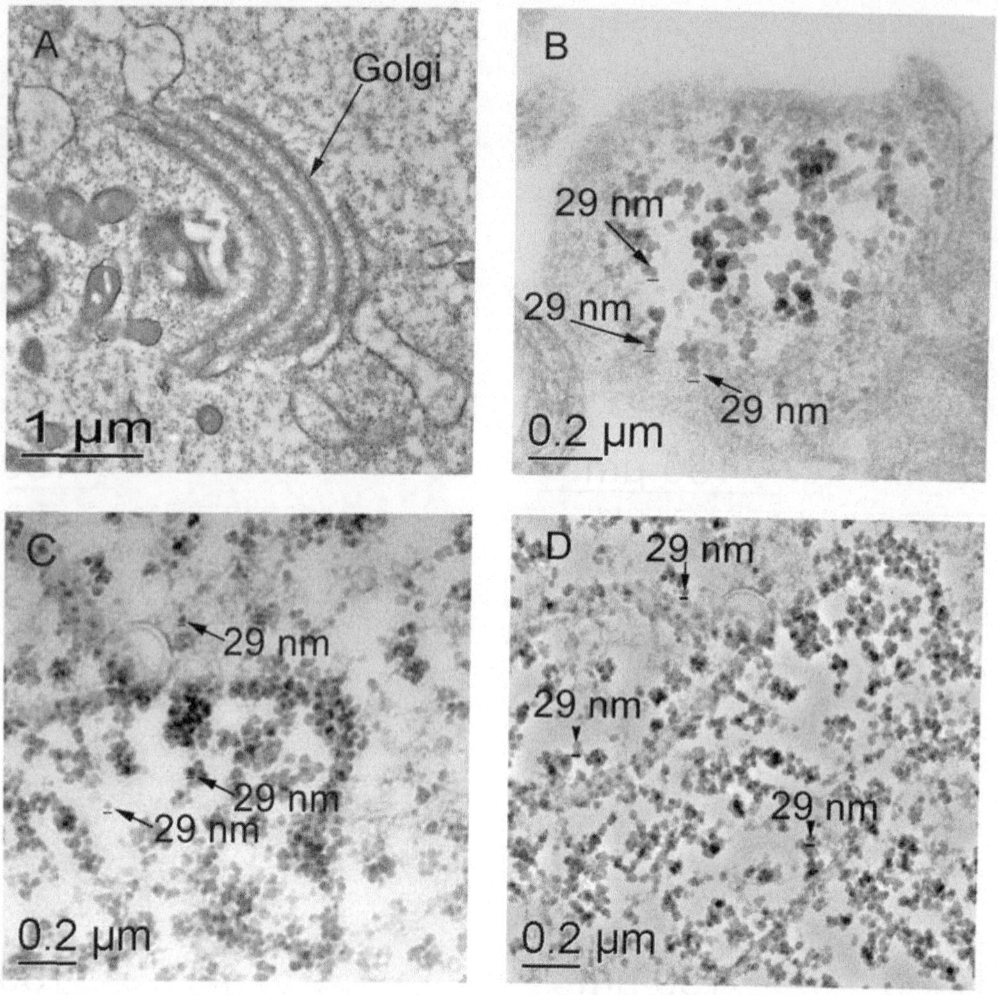


**Figure 3.4 Transmission electron microscopy of uninfected and infected cell cultures from the second infection trial at 48 h pi. A) Uninfected cells showing normal internal membrane organelles. B) Suspected 29-nm particles in cells, viruses from cell culture lysate from the first infection trial (P1). C) Stool sample flag2 (P0). D) Stool sample 149 (P0) showing numerous 29 nm particles and internal rearrangement of membrane-bound organelles (Straub et al. 2007).**

passage in cell culture. Figure 3.5 demonstrates the answers to these questions. Panels A and B show the results of the 3rd passage (P2) of the combined viral stock (containing strains 149, 155 and flag2) in cell culture for "virus-free" samples (generated by passing infectious cell culture lysate through a 10,000 MWCO filter) and the infectious cell culture lysate, respectively. Panels C and D display similar results for the virus free and infectious flag2 stool sample, respectively, and Panel E shows the results for flag2 for its second passage in cell culture. Negative results for RT-PCR and semi-nested PCR confirmed our observations regarding the lack of pathology exhibited in Panels A and C, while RT-PCR and semi-nested PCR were positive for all tissue aggregates receiving virus (Panels B, D, and E).

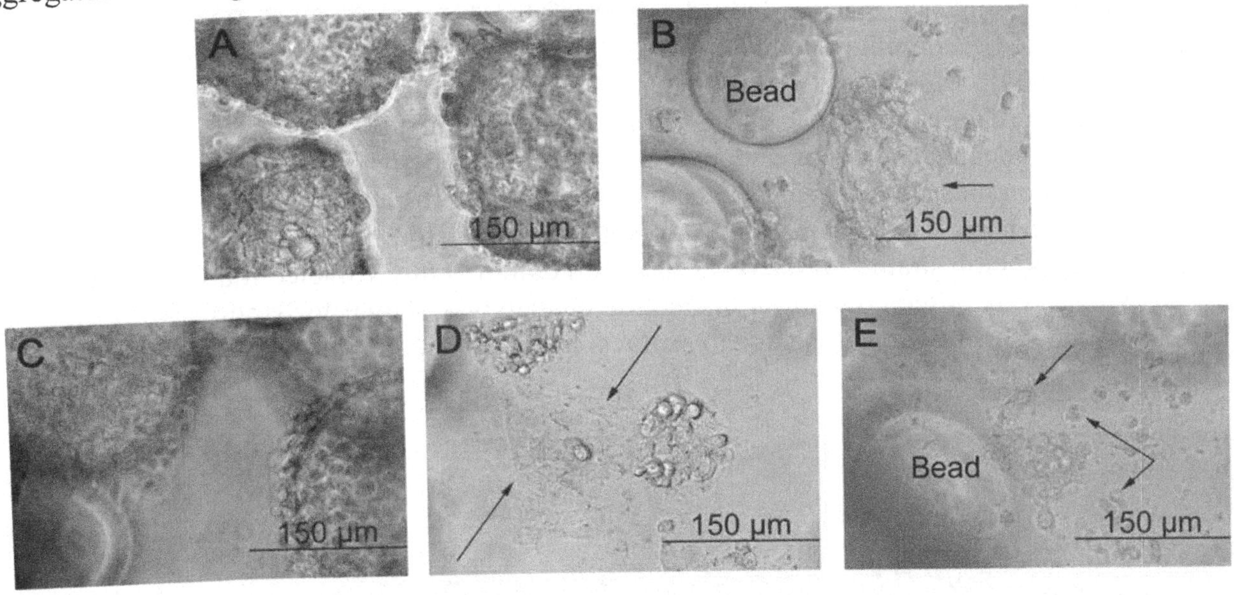


**Figure 3.5 Cytopathic effect results from the third infectivity trial at 24 hours pi. A) Virus-free control of B) combined viral stock lysate from second passage experiment (second infectivity trial, P1), which was used to infect naive cells (P2). C) Virus-free control of the flag2 stool sample. D) Corresponding infection with the flag2 stool sample (P0). E) Flag2 in cell culture (P1). Cells in Panels B, D, and E were confirmed as positive for norovirus by reverse transcription-PCR (RT-PCR) and seminested PCR. Cells in uninfected controls were negative for norovirus by both RT-PCR and nested PCR. Arrows indicate cells exhibiting unusual pathology (Straub et al. 2007).**

## FOURTH INFECTION TRIAL DECEMBER 2005.

The fourth infectivity trial investigated three questions. First, if replication were not occurring, then we would expect that after 5-6 passages through cell culture, there would be a lack of pathology and lack of nucleic acid detection in the later passages. Our experimental design was such that the 5th and 6th passages represented an effective dilution of 1:100,000 and 1:1,000,000, respectively. Moreover, both stool sample stocks were negative by RT-PCR, but positive by semi-nested PCR indicating very low relative viral titer. Second, we wanted to definitively determine whether antigenically distinct isolates could cause infection in cell culture. Sample 155 was initially sequenced and characterized to be a genogroup I isolate, and the flag2 sample was initially characterized to be a genogroup II isolate. Finally we wanted to know whether the nucleic acid sequences isolated from cell culture were changing, perhaps indicating attenuation in cell culture.

For this trial, we used genogroup specific molecular beacons designed around the highly conserved ORF1-ORF2 junction (Kageyama et al. 2003) to target intracellular norovirus RNA. Figure 3.6 displays confocal laser scanning microscopy images of uninfected tissue aggregates after 4 passages in cell culture, and tissue aggregates infected with samples 155 and flag2 after 6 passages in cell culture (Panels A – C, respectively). Lack of a specific signal is observed in uninfected tissue aggregates, whereas abundant molecular beacon signal is observed for both samples, especially at the 5th passage. If the viruses were truly being diluted to extinction, we would not expect to see specific hybridization of the beacon to viral RNA, or if so, the distribution of the beacon might be localized to a few scattered spots as reported by Duizer et al. (Duizer et al. 2004). Moreover, CPE, similar to the previous three infection trials, was observed in all 6 passages (data not shown). These results are thus indicative that 1) viral infectivity and replication is not dependent on diverse antigenic differences between isolates, and that 2) some level of replication must be occurring, given the robust signal from the molecular beacon assay. We further sequenced RT- PCR products from the original stool sample that contained strains 155 and flag2 and both strains from passage 3 in cell culture. Only 1 nucleotide substitution for passage 3 flag2 was observed, and no nucleotide change was shown for strain 155 (Table 3.2). This nucleotide substitution may be due to artifacts in the sequencing process.

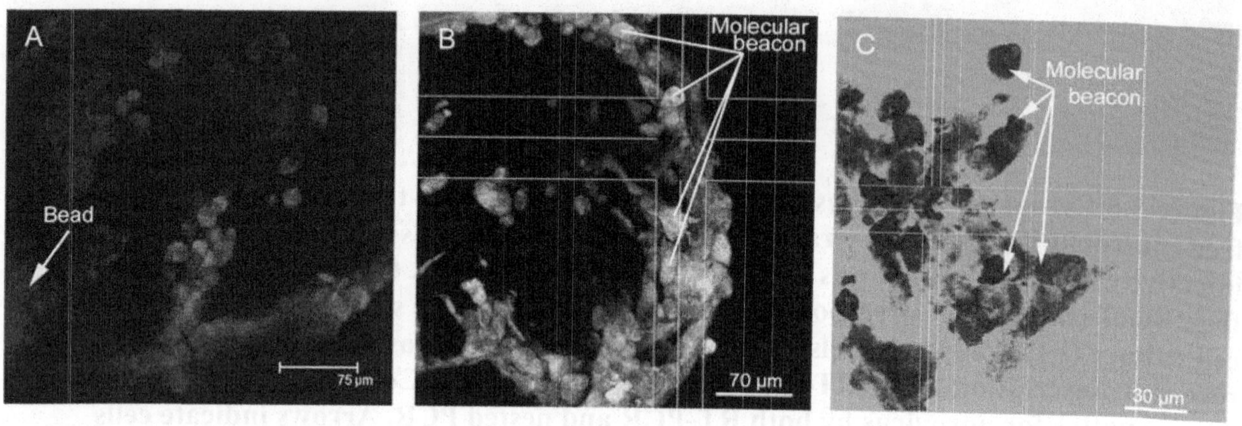


**Figure 3.6 Deconvolved confocal laser scanning micrographs of the molecular beacon fluorescence *in situ* hybridization assay, demonstrating viral infectivity of a genogroup I virus (Sample 155) and genogroup II virus (flag2) at 24 hours postinfection. A) Typical response for uninfected cells, no molecular beacon observed. B) Sample 155, P5 in cell culture. C) Sample flag2, P5 in cell culture (Straub et al. 2007).**

**Table 3.2**

**Stool vs. cell culture sequencing data of the RNA dependent RNA polymerase gene. The underlined and bold highlighted base indicates the single nucleotide substitution for flag2.**

```
155preTC       AATTATGACAGAGTCCTTCTCCATC
155postTC      AATTATGACAGAGTCCTTCTCCATC
               *************************

155preTC       ATGTCACGCCTTACGGCCTCTCCAGAACTAGCTGAGGTTGTAGCCCAGGATTTATTAGCA
155postTC      ATGTCACGCCTTACGGCCTCTCCAGAACTAGCTGAGGTTGTAGCCCAGGATTTATTAGCA
               ************************************************************

155preTC       CCATCCGAGATGGATGTGGGCGACTATGTTATAAGGGTCAAAGAAGGCCTACCATCAGGA
155postTC      CCATCCGAGATGGATGTGGGCGACTATGTTATAAGGGTCAAAGAAGGCCTACCATCAGGA
               ************************************************************

155preTC       TTTCCCTGCACTTCCCAAGTGAATAGCATAAATCACTGGATAATCACCCTCTGTGCATTG
155postTC      TTTCCCTGCACTTCCCAAGTGAATAGCATAAATCACTGGATAATCACCCTCTGTGCATTG
               ************************************************************

155preTC       TCTGAAGCTACTGGCTTATCACCTGATGTGGTACAGTCCATGTCATACTTCTC
155postTC      TCTGAAGCTACTGGCTTATCACCTGATGTGGTACAGTCCATGTCATACTTCTC
               *****************************************************

flag2 cell     AACACAACAGAGAGCCGTGTCGGCAGCAGC
flag2 stool    AACACAACAGAGAGCCGTGTTGGCAGCAGC
               ******************** *********

flag2 cell     TCTAGAGATCATGGTTAAGTTCTCCTCAGAACCACATTTGGCTCAGGTAG
flag2 stool    TCTAGAGATCATGGTTAAGTTCTCCTCAGAACCACATTTGGCTCAGGTAG
               **************************************************

flag2 cell     TCGCAGAAGACCTTCTTTCTCCTAGCATGGTGGATGTGGGTGACTTCACA
flag2 stool    TCGCAGAAGACCTTCTTTCTCCTAGCATGGTGGATGTGGGTGACTTCACA
               **************************************************

flag2 cell     ATATCAATCAATGAGGGTCTCCCCTCTGGGGTACCCTGCACCTCCCAATG
flag2 stool    ATATCAATCAATGAGGGTCTCCCCTCTGGGGTACCCTGCACCTCCCAATG
               **************************************************

flag2 cell     GAACTCCATCGCCCACTGGCTCCTCACTCTCTGCGCGCTCTCTGAAGTCA
flag2 stool    GAACTCCATCGCCCACTGGCTCCTCACTCTCTGCGCGCTCTCTGAAGTCA
               **************************************************

flag2 cell     CAAATCT
flag2 stool    CAAATCT
               *******
```

# CHAPTER 4 ADDITIONAL RESULTS DEMONTRATING NoV INFECTION

## INTRODUCTION

Our work published in *Emerging and Infectious Diseases* provided a significant body of evidence that NoV does infect and replicate in these 3-D INT-407 cells. In addition to these studies, Pacific Northwest National Laboratory (PNNL) has an internal study that is investigating host response to infectious diseases. An integrated approach that examines metabolic volatile organic compound in the head space of culture media, 1-D matrix assisted laser desorption ionization, time of flight mass spectrometry (1-D MALDI), and metabolomics via slow magic angle nuclear magnetic resonance is being tested on the 3-D cell cultures for *Salmonella* infection of 3-D large intestinal cells, *Pseudomonas* infection of lung tissue, and norovirus infection of the 3-D small intestinal model. The question that is being addressed here is whether the discovered host biosignatures are unique, and thus informative for a given pathogen vs. a generalized host response. The data for noroviruses is being shared in this report with an explicit understanding that 1) the signatures we report are only preliminary and not statistically replicated, and 2) we are not making any claims whether these host signatures are unique for norovirus infection of these cells. The principal context for this data is that it provides additional evidence of viral infectivity that needs to be simultaneously examined with other data such as CPE, FISH or RT-PCR. Finally, through optimization of quantitative reverse transcription real-time PCR we provide evidence of true viral replication in cell culture. Results for the biomarkers and quantitative PCR results are discussed below.

## FOURTH INFECTION TRIAL DECEMBER 2005, REVISITED

MB-FISH indicated that productive viral replication was observed after 5 passages through cell culture (Figure 3.6). Supernatants from these cell cultures were submitted for biosignature discovery. Techniques employed for these samples employed GCxGC- TOFMS to find small molecular weight metabolic volatile organic compounds (MVOCs) and 1-D MALDI for host secreted peptides that are < 30,000 daltons. Figure 4.1 displays the results of the MVOC analysis for P4 ($5^{th}$ generation in cell culture) and P5 ($6^{th}$ generation in cell culture). Of particular note, is that regardless of strain (genogroup I (155) and genogroup II (flag2), the host secreted volatile was the same compound. It was observed in both P4 and P5 infected cells, but not in its corresponding uninfected cell culture sample or fresh media that had not come in contact with the cells.

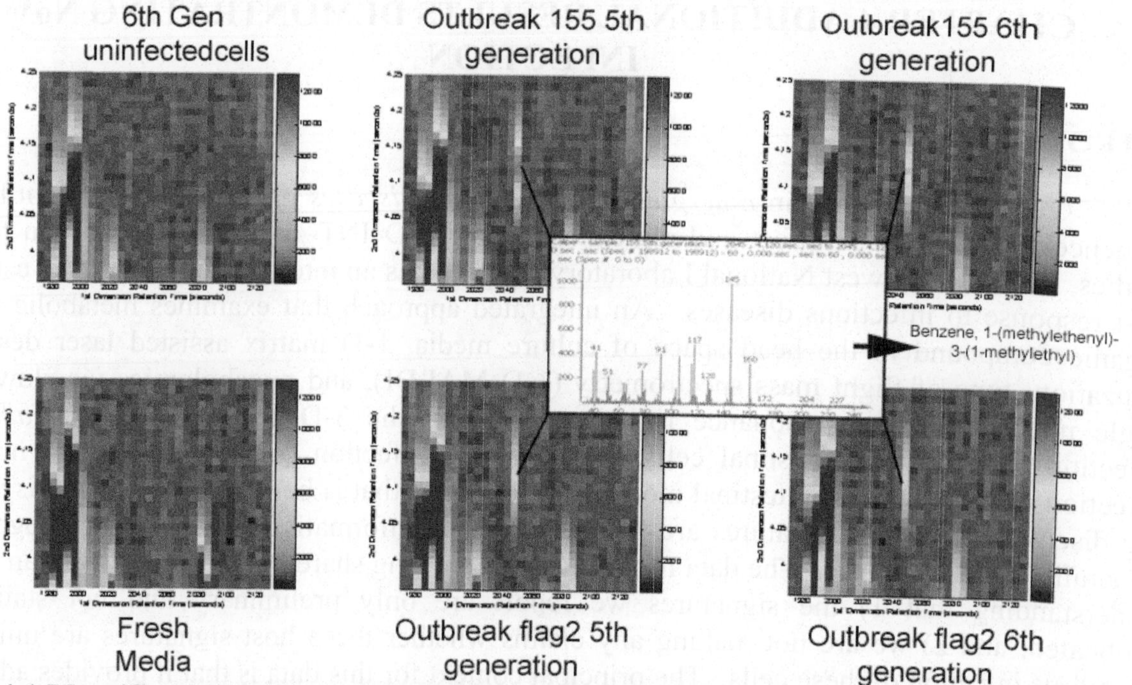


**Figure 4.1 Identified MVOC host biosignature after infection of 3-D INT-407 cells with genogroup I (155) and genogroup II (flag2) noroviruses. This signature appeared at two subsequent passages (P4 = 5th generation and P5 = 6th generation as shown above) but was not genogroup specific. Corresponding uninfected cell cultures and fresh media do not contain this same volatile, or if so, at concentrations significantly less than the infected samples. Figure courtesy of Dr. Jon Wahl, PNNL.**

1-D MALDI analysis of host-secreted peptides revealed that the 3-D INT407 host cells were responding to viral challenge in some way. Figure 4.2 displays the results for both genogroups. Of particular note here is that, unlike the VOC analysis, host response appeared to be strain dependent. Statistical cross validation through 10 iterations revealed an overall score of 97.78% and overall recognition capability of 100%. Basically, this means that a given peptide marker could be identified to either its uninfected, 155, or flag2 infection with a high degree of statistical confidence. However, it is important to note that these signatures are putative peptides, but the actual identity of these peptides (e.g. cytokine, leukotrine, or interferons) has not been determined.

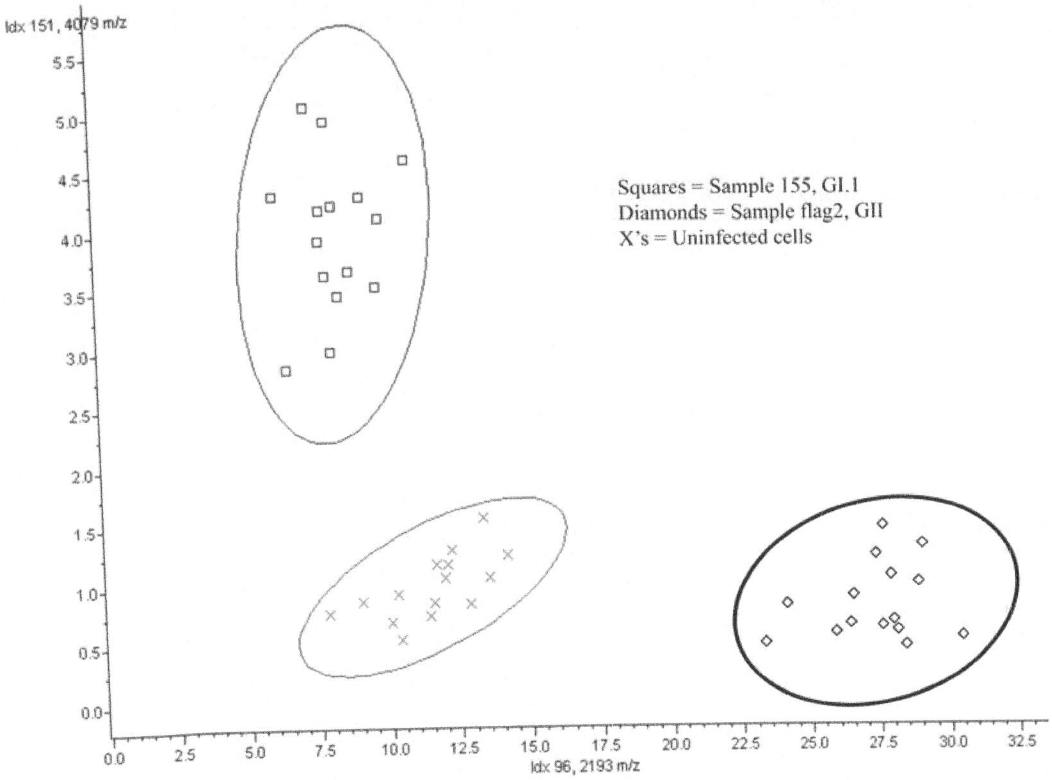


**Figure 4.2 1-D MALDI analysis of host-secreted peptides from norovirus infected cells from P4 and P5 of the 4$^{th}$ infectivity trial. Host-response appeared to be strain dependent in this case. The axes represent the relative peak areas for a given sample and mass/charge ratio. Figure courtesy of Dr. David Wunschel, PNNL.**

Thus the host biosignature discovery work is providing further evidence that norovirus can infect 3-D small intestinal cells. We repeated this work with norovirus infection with 3-D HT-29 (large intestinal cells), but did not get unique signatures. This was not unexpected since our traditional methods for determining viral infectivity (pathology, PCR, etc.) had also failed. However, these findings are also important because it confirms early work that norovirus infection targets the proximal small intestine. In the human challenge studies reported in the 1970's very little involvement of the large intestine was noted (Dolin et al. 1975; Schreiber et al. 1973).

## REVERSE TRANSCRIPTION REAL TIME PCR RESULTS.

Quantitative reverse transcription, real-time PCR was performed to determine the magnitude of propagation in these cell cultures. For this study, we examined norovirus GI.1 stool samples obtained from Emory's human challenge study, and two genogroup II strains, representing a high dose and low dose challenge, respectively. Characterization of these samples is displayed in Table 4.1 below.

**Table 4.1**
**Isolates evaluated by quantitative reverse transcription real-time PCR**

| Sample ID | Date Received | Setting | Strain in GenBank with closest sequence similarity (%, where known) | Genogroup |
|---|---|---|---|---|
| 3-1 | August 2006 | Clinical | M87661 (100%) | NA* |
| 3-9 | August 2006 | Clinical | M87661 (100%) | I.1 |
| 4-1 | August 2006 | Clinical | M87661 (100%) | NA |
| 4-3 | August 2006 | Clinical | M87661 (100%) | I.1 |
| 8G | March 2007 | Cruise ship | AB240184 Norovirus Hu/NV/Hokkaido/284/2004/JP | II |
| 127 | March 2007 | Cruise ship | AY502023 Norovirus genogroup 2 strain Hu/NoV/Farmington Hills | II.4 |
| VS122 | December 2005 | Cruise ship | EU127894097† | II.4 |
| Flag1 | December 2004 | Nursing home | AJ626578 NV/GII/Stockholm/IV1138/2003/SE (97%) | II |

* NA = Pre- clinical challenge with norovirus, viruses were not recovered by real-time PCR or RT-PCR for these controls.

† Our group sequenced the capsid gene for this isolate and submitted to GenBank. The original alignment indicated that this isolate is GII.4 Hunter

Tables 4.2 and 4.3 display the quantitative reverse transcription, real-time PCR infectivity data for supernatant and cell fractions, respectively. Supernatants are defined as the liquid fraction and any cells that had left the intact tissue aggregates. The cell fraction represents viruses recovered from the intact aggregates after 2 washes with 1X PBS. For the results shown in these tables, viruses were minimally diluted 1:100 in tissue culture media, and the initial real-time data was generated from these stocks, and not the original stool sample. Sample 4-3 in the first round of infections was diluted 1:1,000 before being applied to cells. Thus, inhibition due to the stool sample was likely not an issue. This was further confirmed by performing real-time PCR on stool and dilutions of the stool. In all cases, expected approximate 10-fold decreases in titer with increasing dilutions were observed. This indicates that our procedures for viral RNA extraction from stool and cell culture were effective in removing PCR inhibitors.

There are several important points to note regarding propagation of the viruses in cell culture. While it is true that a majority of the outbreaks worldwide today (CDC 2007) are being caused by a highly virulent genogroup II.4 strain, the data shown in Tables 4.2 and 4.3 offer few clues as to why the GII.4 strains may be more virulent. Propagation varied from approximately 10-fold to approximately 1,000 fold regardless of genogroup for the higher dose infections. Virulence may be host driven. If the biomarker results displayed in Figure 4.2 are reproducible, humans respond differently to infection with the two genogroups, and this host response may drive the severity of symptoms observed. Second, unlike a typical enteric virus infection of monolayers where left unchecked, will eventually infect every cell, norovirus infection in the 3-D cell cultures behave much differently. It does so in ways that are (continued on page 35)

**Table 4.2**

**Reverse transcription, real-time PCR data for NoV infections of cell cultures: supernatant fraction**

| | | | Mean copies per mL recovered | | | | |
|---|---|---|---|---|---|---|---|
| Sample ID | Genogroup | Mean copies applied to cell culture* | 24 hrs pi | 48 hrs pi | 72 hrs pi | 96 hrs pi | Setting |
| VS 122 | II.4 | < Limit of detection (LOD)† | < LOD | 1,922 | 7,075 | ND‡ | Cruise ship |
| flag1 | II | < LOD | < LOD | < LOD | < LOD | ND | Nursing home |
| Patient 3-1 | I.1 | < LOD | < LOD | < LOD | < LOD | ND | Clinical pre-challenge |
| Patient 3-9 | I.1 | 80,200 | 989,703 | 36,143 | < LOD | ND | Clinical post challenge |
| Patient 4-1 | I.1 | < LOD | < LOD | < LOD | < LOD | ND | Clinical pre-challenge |
| Patient 4-3 | I.1 | 6,960 | 6,840,255 | 165,335 | 396,062 | ND | Clinical post challenge |
| Patient 4-1§ | I.1 | < LOD | < LOD | < LOD | < LOD | ND | Clinical pre-challenge |
| Patient 4-3§ | I.1 | 96,500 | 21,685,800 | 5,536,800 | 1,007,390 | 422,950 | Clinical post challenge |
| 127** | II.4 | <LOD | <LOD | <LOD | <LOD | <LOD | Cruise ship |
| 8G | II | 457,800 | 26,684,300 | 5,767,500 | 1,540,307 | 464,476 | Cruise ship |

* Viruses in stools were diluted in tissue culture media and then applied to cells. This represents the intrinsic copies applied to the cells, prior to being overlain with 1 mL of tissue culture media.

† The LOD for our assay was consistently between 100 and 1,000 reverse transcription real-time PCR copies per mL.

‡ ND = Not performed for this timepoint.

§ This is a repeated experiment with a different batch of cells and on a different date. Note that 3-1 and 4-1 are stool samples from two different patients that are not infected with norovirus (e.g. stool sample negative control).

** Viral replication was observed in the cell fraction at 96 hrs post infection (26,915 copies/mL).

**Table 4.3**
**Reverse transcription, real-time PCR data for NoV infections of cell cultures: cell fraction**

| | | | Mean copies per mL recovered | | | | |
|---|---|---|---|---|---|---|---|
| Sample ID | Genogroup | Mean copies applied to cell culture* | 24 hrs pi | 48 hrs pi | 72 hrs pi | 96 hrs pi | Setting |
| Patient 4-1 | I.1 | < Limit of detection (LOD)† | < LOD | < LOD | < LOD | ND‡ | Clinical pre-challenge |
| Patient 4-3 | I.1 | 96,500 | 703,635 | 507,540 | 87,666 | 52,292 | Clinical post challenge |
| 127§ | II.4 | <LOD | <LOD | <LOD | <LOD | 26,915 | Cruise ship |
| 8G | II | 457,800 | 937,411 | 317,597 | 175,332 | 292,220 | Cruise ship |

* Viruses in stools were diluted in tissue culture media and then applied to cells. This represents the intrinsic copies applied to the cells, prior to being overlain with 1 mL of tissue culture media. Cell culture media is completely changed every 24 hrs. The cell fraction was washed 2X with 1 X PBS, and the supernatant was discarded after each wash. These values represent viral RNA concentration after the cells were washed.

† The LOD for our assay was consistently between 100 and 1,000 reverse transcription real-time PCR copies per mL.

‡ Note that sample 4-1 is a stool samples from a patient that was not infected with norovirus (e.g. stool sample negative control). ND = not performed for this timepoint.

§ The initial stool titer for this sample was 340 copies per mL, but was undetectable after 1:100 dilution in tissue culture media. Note the increase in titer at 96 hrs post infection.

(continued from page 33) physiologically meaningful to actual disease progression in humans. For high dose infections in cell culture, a statistically significant increase in viral titer was observed for both genogroups at 24 hrs post infection. For comparison, for most all outbreak reports, especially those on cruise ships or extremely confined settings, the secondary transmission spike occurs 24 hrs after the index cases. In addition most human symptoms occur within 12-24 hrs exposure followed by rapid recovery 48 – 72 hrs post infection. Likewise our cell culture results showed similar trends in viral clearance, especially since the media was changed every 24 hrs. We hypothesize that this phenomena is an innate immune system response, and not simple viral die off. To support this hypothesis, secreted peptide data for cell cultures infected with 3-9 and 4-3 were clearly differentiated from their respective controls (3-1 and 4-1, respectively, Figure 4.3).

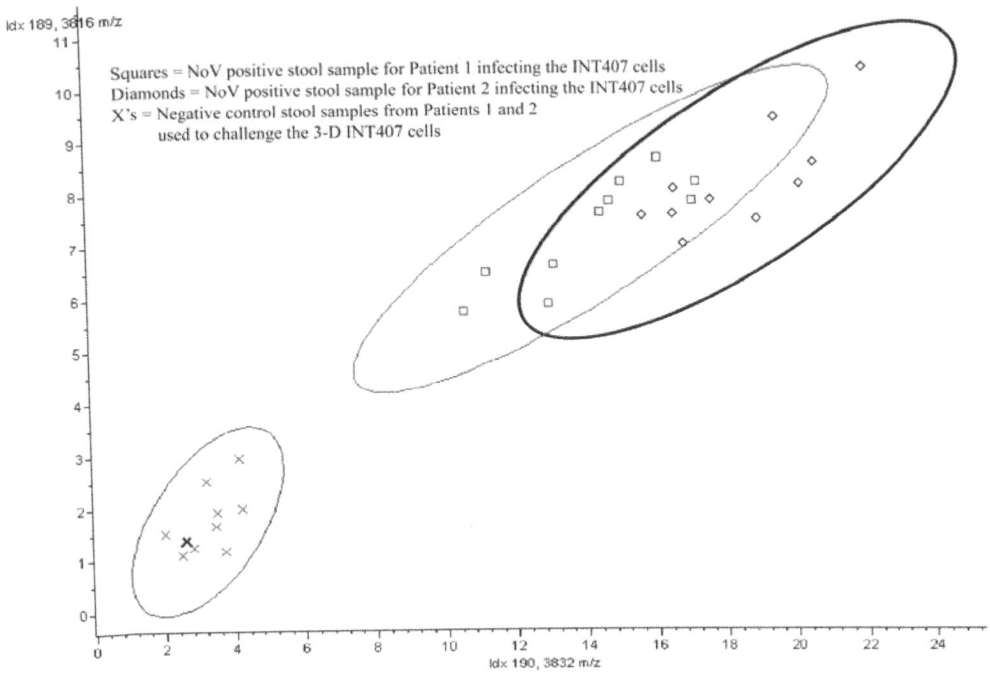


**Figure 4-3 1-D MALDI-MS analysis showing significant differences in host-secreted peptides between uninfected (indicated by the x's) and GI.1 NoV infected cell cultures (indicated by the squares and diamonds (patient samples 3-9, and 4-3)). In this graph, the software picked two mass/charge ratio (m/z) peaks that appeared to be significantly different between uninfected and infected cells, respectively. The axes represent the relative peak areas for a given sample and m/z ratio. Ovals around the indicated peak areas represent the confidence interval for the respective samples. Non-overlapping areas are statistically significant. This data was provided as a courtesy from Dr. David Wunschel Pacific PNNL.**

Low dose infections appear to behave somewhat differently. Both strains VS122 and 127 was classified as a GII.4 strain, but these samples began with estimated titers of between 3 and 300 copies applied to the cells. The estimated multiplicity of infection (ratio of virus particles to cells) was between $1 \times 10^{-6}$ and $1 \times 10^{-4}$ (1 virus interacting with 1,000,000 to 10,000 host cells, respectively). In this case, reverse transcription real-time PCR was negative at 24, 48 hrs. At 72 hrs post infection, VS122 became positive. At 96 hrs 127 also became positive. For both VS122 and 127, the degree of replication was highly significant (average of triplicate infections, estimated at 1,000 to 10,000 fold replication over the baseline titer), especially since the media in both of these infection trials was completely exchanged every 24 hrs. These findings may provide further clues regarding the epidemiology of these viruses. In this scenario, a person (or persons) could be asymptomatic for a period of time (e.g. 1st day on a cruise ship or confined setting). At a critical time point, these people become symptomatic, shedding extremely high numbers. With high secondary transmission rates, this would account for the spike in the number of cases following the initial index cases. More work would need to be conducted to support this hypothesis.

# CHAPTER 5 CONCLUSIONS AND RECOMMENDATIONS

Our primary goal was to develop an infectivity assay for human noroviruses. The evidence provided in the published paper (Straub et al. 2007), and further evidence discussed in this report supports our hypothesis that the human noroviruses can infect and replicate in the *in vitro* 3-D small intestinal epithelial model. Furthermore, this new *in vitro* model appears to mimic both the pathology and innate immune response observed in human challenge studies. Namely,

- For high dose response, appearance of pathology in cell culture and appearance of symptoms in humans occurs within 24 hrs.
- Sub-cellular pathology appears similar in both the 3-D INT 407 and human biopsy data.
- For these high dose challenges, maximum viral shedding is correlated with appearance of symptoms, but decreases over time. While we did not specifically show this data, over the time course of an infection, we observed that the aggregates began to repair themselves. We confirmed this finding by dissociating the remaining cells from the beads, and transferring them to a T-75 flask. These surviving cells formed healthy monolayers. When these monolayers were then transferred back into the RWV bioreactor, healthy aggregates, showing no signs of pathology or viral infection (as confirmed by both reverse transcription PCR and reverse transcription real-time PCR) was observed. When these new 3-D aggregates were challenged with norovirus, they too became infected and viral replication occurred. ***This is actually the data displayed in Tables 4.2 and 4.3***. This could be expected because these 3-D tissue assemblies do not have an adaptive immune response, and innate immune responses are generally short lived and localized.

## RECOMMENDATIONS FROM OUR FINDINGS TO THE WATER INDUSTRY

While the results point to successful infection of the 3-D INT407 cells with human noroviruses, our results are not what we expected in terms of the development of a standardized assay for the water industry. Namely, we were expecting either infection with no productive viral replication as we observe with *Cryptosporidium parvum* (Rochelle et al. 2002) or unchecked propagation of the viruses as we observe for our traditional enteric virus assays on monolayers. For this assay, we observed productive viral replication followed by attempted clearing of the virus in this system. If our results are confirmed, this has several implications for both disinfection studies and standardized monitoring methods. Namely:

- If we need to propagate the virus to high titer to perform disinfection studies, we will need to first confirm early replication followed by viral clearing in the 3-D cell culture system. Optimization of viral replication may require identification of the cytokines, chemokines, leukotreines, and interferons, and neutralizing these compounds prior to, or during viral infection.
- For standardized methods, we will likely need multiple lines of evidence that viral infection has occurred. This may include a combination of gross pathological examination, fluorescence *in situ* hybridization, and quantitative reverse transcription

real-time PCR assays. However, these assays should be conducted over a 24 to 96 hr to capture both the low and high dose response to infection.

Finally, we strongly recommend that water utility laboratories and state and federal agencies that wish to learn these new methods receive the appropriate training to make this assay work for them. Most laboratories with cell culture capabilities can learn these methods, but should realize that these methods are not just about having the cells grow on a bead. Complex physiological processes need to occur to have the cells differentiate to the level that they become permissive to norovirus infection. This includes specialized media requirements and the care and feeding of these 3-D cell cultures. Finally, QC tests should be performed to insure that the cells that are being infected display the desired characteristics described in this report.

# APPENDIX A: PERMISSION TO REPRODUCE SECTIONS OF STRAUB ET AL. 2007

Dear Dr. Straub:

Emerging Infectious Diseases is published by the Centers for Disease Control and Prevention, a federal agency. Therefore, all our published materials are in the public domain and can be used without permission. Proper citation is required.

If you are requesting high resolution images, please contact Ann Jordan at ajordan@cdc.gov <mailto:ajordan@cdc.gov> , and she will generate the images for you.

Thank you for your interest in our journal.

Best regards,
Susanne Justice
Journal Administrator

P. Susanne Justice
Emerging Infectious Diseases Journal
NCPDCID/CCID/CDC
1600 Clifton Rd., M/S D-61
Phone: 404.639.1759
EID Journal Office: 404.639.1960
Fax: 404.639.1954
Celebrating CDC's First 60 Years

## REFERENCES


Abbott, A. 2003. Cell Culture: Biology's New Dimension. *Nature* 424: 870-872.

Asanaka, M., R. L. Atmar, V. Ruvolo, S. E. Crawford, F. H. Neill and M. K. Estes 2005. Replication and Packaging of Norwalk Virus Rna in Cultured Cells. *Proc. Natl. Acad. Sci.* 102(29): 10327-10332.

Atmar, R. L. and M. K. Estes 2001. Diagnosis of Noncultivatable Gastroenteritis Viruses, the Human Caliciviruses. *Clin. Microbiol. Rev.* 14(1): 15-37.

Cannon, J. L., E. Papafragkou, G. W. Park, J. Osborne, L. A. Jaykus and J. Vinje 2006. Surrogates for the Study of Norovirus Stability and Inactivation in the Environment: A Comparison of Murine Norovirus and Feline Calicivirus. *J. Food Prot.* 69(11): 2761-2765.

Carterson, A. J., K. Honer zu Bentrup, Ott. C.M., M. S. Clarke, D. L. Pierson, C. R. Vanderburg, K. L. Buchanan, C. A. Nickerson and M. J. Schurr 2005. A549 Lung Epithelial Cells Grown as Three-Dimensional Aggregates: Alternative Tissue Culture Model for *Psuedomonas Aeruginosa*. *Infect. Immun.* 73(2): 1129-140.

Carvalho, H. M., L. D. Teel, G. Goping and A. D. O'Brien 2005. A Three Dimensional Tissue Culture Model for the Study of Attach and Efface Lesion Formation by Enteropathogenic and Enterohaemorrhagic *Esherichia Coli*. *Cell Microbiol.* 7: 1771-1781.

Caul, E. O. and H. Appleton 1982. The Electron Microscopical and Physical Characteristics of Small Round Human Fecal Viruses: An Interim Classification Scheme. *J. Med. Virol.* 9(4): 257-265.

CDC 2007. Norovirus Activity - United States, 2006-2007. *MMWR* 56(33): 842-846.

CDC (MMWR 2001; 50 (No. RR-9)). "Norwalk-Like Viruses": Public Health Consequences and Outbreak Management. Atlanta, GA, United States Centers for Disease Control and Prevention: 1-24.

Chang, K.-O., S. V. Sosnovtsev, G. Belliot, A. D. King and K. Y. Green 2006. Stable Expression of a Norwalk Virus Rna Replicon in a Human Hepatoma Cell Line. *Virology* 353(2): 463-473.

Cheetham, S., M. Souza, R. McGregor, T. Meulia, Q. Wang and L. J. Saif 2007. Binding Patterns of Human Norovirus-Like Particles to Buccal and Intestinal Tissue of Gnotobiotic Pigs in Relation to a/H Histo-Blood Group Antigen Expression. *J. Virol.* Jan 10: Epub ahead of print.

Cheetham, S., M. Souza, T. Meulia, S. Grimes, Han M-G. and L. J. Saif 2006. Pathogenesis of a Genogroup Ii Human Norovirus in Gnotobiotic Pigs. *J. Virol.* 80(21): 10372-10381.

de Wit, M. A., M. P. Koopmans, L. M. Kortbeek, W. J. Wannet, J. Vinje, F. van Leusden, A. I. Bartelds and Y. T. van Duynhoven 2001. Sensor, a Population-Based Cohort Study on Gastroenteritis in the Netherlands: Incidence and Etiology. *Am. J. Epidemiology* 154: 666-674.

Dolin, R., A. G. Levy, R. G. Wyatt, T. S. Thornhill and J. D. Gardner 1975. Viral Gastroenteritis Induced by the Hawaii Agent. Jejunal Histopathology and Serologic Response. *Am. J. Med.* 59: 761-768.

Duizer, E., K. G. Schwab, F. H. Neill, R. L. Atmar, M. P. G. Koopmans and M. K. Estes 2004. Laboratory Efforts to Cultivate Noroviruses. *J. Gen. Virol.* 85: 79-87.

Fernandez-Vega, V., S. V. Sosnovtsev, G. Belliot, A. D. King, T. Mitra, A. Gorbalenya and K. Y. Green 2004. Norwalk Virus N-Terminal Nonstructural Protein Is Associated with Disassembly of the Golgi Complex in Transfected Cells. *J. Virol.* 78(9): 4827-4837.

Fout, G. S., B. C. Martinson, M. W. N. Moyer and D. R. Dahling 2003. A Multiplex Reverse Transcription-Pcr Method for Detection of Human Enteric Viruses in Groundwater. *Appl. Environ. Microbiol.* 69(6): 3158-3164.

Green, K. Y., T. Ando, M. S. Balayan, I. N. Clarke, M. K. Estes, D. O. Matson, S. Nakata, J. D. Neill, M. J. Studdert and H. J. Thiel 2000. *Caliciviridae*. Virus Taxonomy: 7th Report of the International Committee on Taxonomy of Viruses. C. M. van Regenmortel, C. M. Fauquet, D. H. L. Bishopet al. Orlando, FL, Academic Press.

Green, K. Y., A. Mory, M. H. Fogg, A. Weisberg, G. Belliot, M. Wagner, T. Mitra, E. Ehrenfeld, C. E. Cameron and S. V. Sosnovtsev 2002. Isolation of Enzymatically Active Replication Complexes from Feline Calicivirus-Infected Cells. *J. Virol.* 76(17): 8582-8595.

Hale, A., K. Mattick, D. Lewis, M. Estes, X. Jiang, J. Green, R. Elgin and D. Brown 2000. Distinct Epidemiological Patterns of Norwalk -Like Virus Infection. *J. Med. Virol.* 62: 99-103.

Hale, A. D., T. N. Tanaka, N. Kitamoto, M. Ciarlet, X. Jiang, N. Takeda, D. W. G. Brown and M. K. Estes 2000. Identification of an Epitope Common to Genogroup 1 "Norwalk-Like Viruses". *J. Clin. Microbiol.* 38(4): 1656-1660.

Harrington, P. R., L. Lindesmith, B. Yount, C. L. Moe and R. S. Baric 2002. Binding of Norwalk Virus-Like Particles to Abh Histo-Blood Group Antigens Is Blocked by Antisera from Infected Human Volunteers or Experimentally Vaccinated Mice. *J. Virol.* 76(23): 12335-12343.

Harrington, P. R., J. Vinje, C. L. Moe and R. S. Baric 2004. Norovirus Capture with Histo-Blood Group Antigens Reveals Novel Virus-Ligand Interactions. *J. Virol.* 78(6): 3035-3045.

Honer zu Bentrup, K., R. Ramamurthy, M. Ott, K. Emami, M. Nelman-Gonzalez, J. W. Wilson, E. G. Richter, T. Goodwin, D. Pierson, S. Alexander, N. Pellis, K. Buchanan and C. A. Nickerson 2006. 3-D Organotypic Models of Human Colonic Epithelium to Study the Early Stages of Enteric Salmonellosis. *Microbes and Infection* 8(7): 1813-1825.

Hutson, A. M., R. L. Atmar, D. M. Marcus and M. K. Estes 2003. Norwalk Virus-Like Particle Hemagglutination by Binding to H Histo-Blood Group Antigens. *J. Virol.* 77(1): 405-415.

Jothikumar, N., J. A. Lowther, K. Henshilwood, D. N. Lees, V. R. Hill and J. Vinje 2005. Rapid and Sensitive Detection of Noroviruses by Using Taqman-Based One-Step Reverse Transcription-Pcr Assays and Application to Naturally Contaminated Shellfish Samples. *Appl. Environ. Microbiol.* 71(4): 1870-1875.

Kageyama, T., S. Kojima, M. Shinohara, K. Uchida, S. Fukushi, F. B. Hoshino, N. Takeda and K. Katayama 2003. Broadly Reactive, Highly Sensitive Assay for Norwalk-Like Viruses Based on Real-Time Quantitative Reverse Transcription-Pcr. *J. Clin. Microbiol.* 41(4): 1548-1557.

Kapikian, A. Z., R. G. Wyatt, R. Dolin, T. S. Thornhill, A. R. Kalica and R. M. Chanock 1972. Visualization by Immune Electron Microscopy of a 27-Nm Particle Associated with Acute Infections of Nonbacterial Gastroenteritis. *J. Virol.* 10(5): 1075-1081.

Lamarca, H. L., Ott. C.M., K. Honer zu Bentrup, C. L. Leblanc, D. L. Pierson, A. B. Nelson, A. B. Scandurro, G. S. Whitley, C. A. Nickerson and C. A. Morris 2005. Three-Dimensional Growth of Extravillous Cytotrophoblasts Promotes Differentiation and Invasion. *Placenta* 26(10): 709-720.

Lelkes, P. L., E. Ramos, V. V. Nikolaychik, D. M. Wankowski, B. R. Unsworth and T. J. Goodwin 1997. Gtsf-2: A New Versatile Cell Culture Medium for Diverse Normal and Transformed Mammalian Cells. *In Vitro Cell Dev. Biol. - Animal* 33: 344-351.

Lindesmith, L., C. Moe, S. Marionneau, N. Ruvoen, X. Jiang, L. Lindblad, P. Stewart, J. LePendu and R. Baric 2003. Human Susceptibility and Resistance to Norwalk Virus Infection. *Nat. Med.* 9(5): 548-553.

Long, J. P. and J. H. Hughes 2001. Epstein-Barr Virus Latently Infected Cells Are Selectively Deleted in Simulated Microgravity Cultures. *In Vitro Cell. Dev. Biol. Anim.* 37: 223-230.

Long, J. P., S. Pierson and J. H. Hughes 1999. Suppression of Epstein-Barr Virus Reactivation in Lymphobalstoid Cells Cultured in Simulated Microgravity. *In Vitro Cell. Dev. Biol.* 35: 49-54.

Malik, Y., S. Maherchandani, P. Allwood and S. Goyal 2005. Evaluation of Animal Origin Cell Cultures for in Vitro Cultivation of Noroviruses. *J. Appl. Res.* 5(312-317).

Marionneau, S., N. Ruvoen, B. Le Moullac-Vaidye, M. Clement, A. Cailleau-Thomas, G. Ruiz-Palacois, P. Huang, X. Jiang and J. Le Pendu 2002. Norwalk Virus Binds to Histo-Blood Group Antigens Present on Gastroduodenal Epithelial Cells of Secretor Individuals. *Gastroenterology* 122(1967-1977).

Nickerson, C. A., T. J. Goodwin, J. Terlonge, C. M. Ott, K. L. Buchanan, W. C. Uicker, K. Emami, C. L. LeBlanc, R. Ramamurthy, M. S. Clarke, C. R. Vanderburg, T. Hammond and D. L. Pierson 2001. Three Dimensional Tissue Assemblies: Novel Models for the Study of *Salmonella Enterica* Serovar Typhimurium Pathogenesis. *Infect. Immun.* 69(11): 7106-7120.

Nickerson, C. A., K. Honer zu Bentrup and C. M. Ott 2005. Three-Dimensional Cell Culture Models for Drug Discovery and Infectious Disease. *Bioforum Europe* 6: 34-36.

Nickerson, C. A. and M. Ott 2004. A New Dimension in Modeling Infectious Disease. *ASM News* 70(4): 169-175.

Parwani, A. V., W. T. Flynn, K. L. Gadfield and L. J. Saif 1991. Serial Propagation of Porcine Enteric Caicivirus in a Continuous Cell Line. Effect of Medium Supplementation with Intestinal Contents or Enzymes. *Arch. Virol.* 120: 115-122.

Rochelle, P. A., M. M. Marshall, J. R. Mead, A. M. Johnson, D. G. Korich, J. S. Rosen and R. DeLeon 2002. Comparison of in Vitro Cell Culture and Mouse Assay for Measuring Infectivity of *Cryptosporidium Parvum*. *Appl. Environ. Microbiol.* 68(8): 3809-3817.

Rockx, B. H. G., W. M. J. M. Bogers, J. L. Heeney, G. van Amerongen and M. P. G. Koopmans 2005. Experimental Norovirus Infections in Non-Human Primates. *J. Med. Virol.* 75: 313-320.

Schreiber, D. S., N. R. Blacklow and J. S. Trier 1973. The Mucosal Lesion of the Proximal Small Intestine in Acute Infectious Nonbacterial Gastroenteritis. *N. Engl. J. Med.* 288: 1318-1323.

Straub, T. M., K. Honer zu Bentrup, P. Orosz-Coghlan, A. Dohnalkova, B. K. Mayer, R. Bartholomew, C. O. Valdez, C. J. Bruckner-Lea, C. P. Gerba, M. Abbaszadegan and C. A. Nickerson 2007. In Vitro Cell Culture Assay for Human Noroviruses. *Emerg. Infect. Dis.* 13(3): 396-403.

Subekti, D. S., P. Tjaniadi, M. Lesmana, J. McArdle, D. Iskandriati, I. N. Budiarsa, P. Walujo, I. H. Suparto, I. Winoto, J. R. Campbell, K. R. Porter, D. Sajuthi, A. A. Ansari and B. A. Oyofo 2002. Experimental Infection of *Macaca Nemestrina* with a Toronto Norwalk-Like Virus of Epidemic Viral Gastroenteritis. *J. Med. Virol.* 66: 400-406.

Vinje, J., R. A. Hamidjaja and M. D. Sobsey 2004. Development and Application of a Capsid Vp1 (Regiond) Based Reverse Transcription Pcr Assay for Genotyping of Genogroup I and Ii Noroviruses. *J. Virol. Meth.* 116: 109-117.

White, L. J., J. M. Ball, M. E. Hardy, T. N. Tanaka, N. Kitamoto and M. K. Estes 1996. Attachment and Entry of Recombinant Norwalk Virus Capsids to Cultured Human and Animal Cell Lines. *J. Virol.* 70(10): 6589-6597.

Wobus, C. E., S. E. Karst, L. B. Thackray, K.-O. Chang, S. V. Sosnovtsev, G. Belliot, J. M. Mackenzie, K. Y. Green and H. W. I. Virgin 2004. Replication of Norovirus in Cell Culture Reveals a Tropism for Dendritic Cells and Macrophages. *PLOS Biology* 2(12): 2076-2084.

Wobus, C. E., L. B. Thackray and H. W. I. Virgin 2006. Murine Norovirus: A Model System to Study Norovirus Biology and Pathogenesis. *J. Virol.* 80(11): 5104-5112.

Zheng, D.-P., T. Ando, R. L. Frankhauser, R. S. Beard, R. I. Glass and S. S. Monroe 2006. Norovirus Classification and Proposed Strain Nomenclature. *Virology* 346: 312-323.

# LIST OF ABBREVIATIONS

| | |
|---|---|
| ATCC | American Type Culture Collection |
| AWWA | American Water Works Association |
| AwwaRF | Awwa Research Foundation |
| | |
| ca. | Single measurement |
| $^{o}C$ | Degrees Celcius |
| CCL | Candidate Contaminant List |
| CCL-6 | Cell Culture Line – 6, refers to the ATCC part number for the INT 407 cells |
| CDC | United States Centers for Disease Control and Prevention |
| CPE | Cytopathic Effect |
| | |
| DMSO | Dimethylsulfoxide |
| DNA | Deoxyribonucleic Acid; cDNA is complimentary DNA that is produced in a reverse transcription reaction to enable PCR of RNA based noroviruses. |
| dNTP | Deoxyribonucelic acid nucleotide triphosphate |
| | |
| FISH | Fluorescent *In situ* Hybridization |
| | |
| GC x GC TOFMS | Gas chromatography x gas chromatography time-of-flight-mass- spectrometry |
| GTSF-2 | Goodwin's Triple Sugar Formulation – 2, the cell culture media used in this work |
| | |
| HARV | High Aspect Rotating Vessel |
| hr | Hour |
| | |
| INT | Intestinal 407 cells from ATCC |
| | |
| LM | Light Microscopy |
| | |
| MALDI | Matrix Assisted Laser Desorption Ionization Mass Spectrometry |
| MB | Molecular Beacon |
| min | minute |
| MMWR | Morbidity Mortality Weekly Report published by the CDC |
| MWCO | Molecular Weight Cutoff Filter |
| | |
| NoV | Human noroviruses; rNoV recombinant norovirus particles. These are capsids expressed in other vectors such as insects and plants. |
| nt | nucleotide; usually referred to in this context: 3101 nt plasmid refers to the the number of nucleotide bases in the plasmid. |
| | |
| ORF | Open Reading Frame |
| | |
| PBS | Phosphate Buffered Saline; D-PBS is Dulbecco's Phosphate Buffered Saline |
| PCR | Polymerase Chain Reaction |
| pi | post infection |

| | |
|---|---|
| PNNL | Pacific Northwest National Laboratory |
| RNA | Ribonucleic Acid |
| RPM | Revolutions per Minute |
| RWV | Rotating Wall Vessel Bioreactor |
| RT-PCR | In this text, RT-PCR means Reverse Transcription Polymerase Chain Reaction. If real time PCR is performed, then it is spelled out: reverse transcription real time PCR, to avoid confusion. |
| QC | Quality Control |
| s | second(s) |
| STLV | Slow Turning Lateral Vessel |
| $TCID_{50}$ | Tissue Culture Infectious Dose 50; dilution at which 50% of the replicates show signs of viral infection. |
| TEM | Transmission Electron Microscopy |
| Tx-100 | Triton X-100 |
| USEPA | United States Environmental Protection Agency |
| UV | Ultraviolet |
| VOC | Volatile Organic Compounds |

IWA Publishing's authorised EU representative for General Product Safety Regulations is Diane D'Arras, 15 rue Duret, 75116 Paris, France, e-mail: safety@iwap.co.uk.

Printed and bound by CPI Group (UK) Ltd, Croydon, CR0 4YY
17/09/2026
02218810-0007